Food in a Just World

We would like to dedicate this book to our children and all future generations of humans and other animals who will inherit this planet. May we leave a legacy of resilience and compassion. All my relations.

Food in a Just World

Compassionate Eating in a Time of Climate Change

TRACEY HARRIS & TERRY GIBBS

polity

Copyright © Tracey Harris and Terry Gibbs 2024

The right of Tracey Harris and Terry Gibbs to be identified as Authors of this Work has been asserted in accordance with the UK Copyright, Designs and Patents Act 1988.

First published in 2024 by Polity Press

Polity Press
65 Bridge Street
Cambridge CB2 1UR, UK

Polity Press
111 River Street
Hoboken, NJ 07030, USA

All rights reserved. Except for the quotation of short passages for the purpose of criticism and review, no part of this publication may be reproduced, stored in a retrieval system or transmitted, in any form or by any means, electronic, mechanical, photocopying, recording or otherwise, without the prior permission of the publisher.

ISBN-13: 978-1-5095-5401-0
ISBN-13: 978-1-5095-5402-7(pb)

A catalogue record for this book is available from the British Library.

Library of Congress Control Number: 2023936995

Typeset in 11 on 13pt Scala
by Fakenham Prepress Solutions, Fakenham, Norfolk NR21 8NL
Printed and bound in Great Britain by TJ Books Ltd, Padstow, Cornwall

The publisher has used its best endeavours to ensure that the URLs for external websites referred to in this book are correct and active at the time of going to press. However, the publisher has no responsibility for the websites and can make no guarantee that a site will remain live or that the content is or will remain appropriate.

Every effort has been made to trace all copyright holders, but if any have been overlooked the publisher will be pleased to include any necessary credits in any subsequent reprint or edition.

For further information on Polity, visit our website:
politybooks.com

Contents

Detailed Contents vi
Acknowledgments ix
Preface xi

Introduction 1

Chapter 1: Food Justice Needs a Just World: Confronting Structural Violence against Land, Humans, and Nonhuman Animals 12

Chapter 2: Capitalist Dreams and Nightmares: Food Systems, the Animal-Industrial Complex, and Climate Disruption 47

Chapter 3: Working in Hell: Labor in the Industrial Production of Animals as Food 76

Chapter 4: What If We Really Are What We Eat?: Challenging a Colonial-Capitalist Diet 117

Chapter 5: The Upside Down: The Hidden World of Nonhuman Animals as Food 155

Chapter 6: Towards a Compassionate Food System 185

Methods 213
Appendix 218
References 229
Index 254

Detailed Contents

Acknowledgments ix
Preface xi

Introduction 1

Chapter 1: Food Justice Needs a Just World: Confronting Structural Violence against Land, Humans, and Nonhuman Animals 12
   Climate Change and our Relationship to the Land 16
   Earth Democracy and the Democracy of Species 18
   Responding to Structural Violence through "Big C" Compassion 21
   Interdependence as a Framework for Advocacy and Social Protest 25
   Situating Our Relationship to the Land and Other Species within the Law 28
   The Logic of Capital 30
   Corporate Dependency 33
   Towards a Vision of Radical Democracy based in "Big C" Compassion 34
   Neoliberalism and the Myth of "Development" 38
   The Rights of Future Generations 42
   Where Do We Go from Here? 45

Chapter 2: Capitalist Dreams and Nightmares: Food Systems, the Animal-Industrial Complex, and Climate Disruption 47
   Denial as the First Stage of Grief 47
   The Relationship between Human Rights Protection, Peace, and Food Justice 50

From Grief and Denial to Confronting Systems	55
Dominion over the Land	65
The Violence of a Hamburger and Chicken Nuggets ...	67
Recovering from the Green Revolution	71
The Plant-Based Treaty: Moving Toward a Global Shift in Culture?	74

Chapter 3: Working in Hell: Labor in the Industrial Production of Animals as Food — 76

Labor in the Animal-Industrial Complex, COVID-19, and the Logic of Capital	78
The Health and Demographics of Workers	86
Compartmentalization of Human Work Facilitates De-animalization	90
Dirty Work and Slow Violence	92
"That shit will fuck you up for real"	93
The Spillover Effect	98
Deskilling and the "Entanglements of Oppression"	99
Recognizing Animals' Labor	104
Transparency and Beyond: Cultivating Empathy?	108
Resistance and the Beginning of a Just Transition	110

Chapter 4: What If We Really Are What We Eat?: Challenging a Colonial-Capitalist Diet — 117

The Right to Food: Availability, Accessibility, and Adequacy	118
The Issues with Overconsumption and Underconsumption	121
Deconstructing the "Ideal" Body	123
Building Community and Social Infrastructure	126
From Food Insecurity to Food Justice	128
Zoonotic Diseases: How We Treat Other Animals Comes Back to Haunt Us	135
An Incremental Shift to a Compassionate Food Future	146
Food as Political: Resistance Is Not Futile	151
Transparency and Beyond	153

Chapter 5: The Upside Down: The Hidden World of Nonhuman Animals as Food — 155

The Upside Down: Speciesism in Action	156
Why We Don't Know What We Don't Know	158
Individual Abuses and Standard Industry Practices	163

Happy Meat?	175
Turning the World Right-Side Up: Recognizing Nonhuman Animals for Who They Are	181

Chapter 6: Towards a Compassionate Food System — 185
- Radical Democracy? — 186
- How Do We Initiate Positive Change? From Paradigm Shifts to Building Community — 188
- Sorting Out the Politics: Short and Long Term — 198
- The Art of Transparency — 200
- Radical Democracy and the Precautionary Principle — 202
- The Importance of Joy and Gratitude — 204
- Conclusion: Toward the Compassionate World — 209

Methods — 213
Appendix — 218
References — 229
Index — 254

Acknowledgments

First and foremost, we would like to thank all the other species with whom we share this planet and whose presence in the world touches every aspect of our lives. We would also like to thank all those who work tirelessly to defend the rights of other species and those of mother nature herself; those who write and research, those engaged in activism and advocacy, and those engaged in both. We honor those contributing to positive change in their communities often with intense emotional labor and sometimes with great risk. We draw on numerous examples of these dedicated individuals, many cited, and some interviewed for this book. We offer a heartfelt thank you to each of our 28 interview participants who are engaged in incredible work in the places they find themselves. What a great pleasure and learning experience it has been to work with them. We also feel so much gratitude for Elder Albert Marshall who has been a constant source of inspiration. He moves in the world with a resilience and compassion that is almost unfathomable. His friendship, guidance and support have meant the world to us.

There are too many colleagues, friends, and family to name who have supported us in the journey of writing this book, but we would like to acknowledge Zabrina Downton and Eli Quirk, Amber and Sadie Buchanan, Nicky Duenkel and Judy Pratt, Carol Smith, Andrea Donato, Julie Hearn, Carolyn Claire, Kirby Evans, and Pema Chödrön for many discussions, shared meals, tears, laughter, texts, letters, and ongoing support. Our research assistant Ashleigh Long worked tirelessly with us throughout this entire project. Her work was exceptional, and we are so grateful for her support, patience, and openness in this long process. Her support was made possible through generous funding from Cape Breton University. Thank you to Sandi Maxwell who brought her superb editing skills to assist us in polishing our final manuscript. We also extend thanks to our colleagues in the Department of L'nu, Political, and Social Studies and the School of Arts and Social Sciences at Cape Breton University for providing a collegial and supportive context for writing this kind of book.

We are particularly grateful to our fastidious and insightful readers: Erin Gibbs, Garry Leech, Zabrina Downton, Kirby Evans, Stephen Harris, Dana Mount, and Joe Parish who generously carved out space in their busy lives to dedicate to the final stages of this project. Thank you to the reviewers who provided feedback on early drafts of the chapters. We would also like to thank the team at Polity Press, particularly Jonathan Skerrett, Senior Commissioning Editor, for his support and patience, and ultimately for his faith in this rather large and unorthodox project! We also appreciate the work of Karina Jákupsdóttir and Ian Tuttle who have been supporting the editorial process in the background.

We are extremely grateful to Marie Battiste who generously shared her knowledge with us and provided guidance and advice on the use of language and syntax, particularly as they relate to Indigenous Peoples and other equity-deserving groups. Marie also offered thoughtful input on questions to consider with regard to writing on issues related to Indigenous Peoples more generally. We hope we have respectfully and adequately reflected this input and spirit in our book. Any errors or misunderstandings that remain are entirely the responsibility of the authors.

Last, but certainly not least, we must acknowledge the patience, love, and unending support of our beautiful partners, children, and companion animals: Stephen and Olivia Harris, Ashley and Bella; Garry Leech, Owen and Morgan Gibbs Leech, Johan Gallardo, and Shadow: we know we are not easy to live with!

Preface

Our Research Approach

This book draws on many years of research and each of us, as co-authors, brings our own areas of specialization to the broad theme of food in a just world. It is important to be upfront about these experiences and we have attempted throughout this methodological process, and throughout the book itself, to "apply a critical reflexivity to [this] research process" as we "engage in a deep questioning" of ourselves as researchers, and how our own backgrounds, identities, and experiences influence the research project and research findings (Tilley 2016: 13).

Terry Gibbs is a professor of international politics with a focus on social justice and critical globalization studies. While her past research has examined human rights, social justice movements, and democratic practice in the context of economic globalization, she has recently explored the structural implications of systems more broadly by including both nonhuman animals and nature into her framework. In addition to her ongoing solidarity work with social justice and human rights movements in Latin America, she is also engaged locally in Cape Breton/Unama'ki, Canada, with various community projects related to decolonization, the food movement, climate change, and mental health. Tracey Harris is an associate professor of sociology with a focus on critical reflections on the relationship between humans and other animals, sustainable housing, consumer culture, environmental sociology, and qualitative research methodology.

Together, with our colleague Richard Keshen, who teaches philosophy at Cape Breton Universty, we co-founded the Animal Ethics Project, which promotes education and advocacy on issues related to human relationships with other animals. These experiences have influenced and shaped both the practical and theoretical lenses we bring to this research on human relationships with other animals, food security, climate change, and other health and well-being issues.

Part of the reflexivity mentioned involves bringing ourselves into the research by sharing relevant stories, journal entries, and life events that help the reader understand more holistically what this research is all about, and why we have come to see things the way we do (Kovach 2009: 49–50). As Joseph Maxwell makes clear in his book *Qualitative Research Design: An interactive approach* (2005: 79): "Qualitative data are not restricted to the results of specified 'methods' ... you *are* the research instrument in a qualitative study, and your eyes and ears are the tools you use to make sense of what is going on."

In our research this means that autoethnographic tools, such as journaling and systematic notetaking, where we consider our thoughts, feelings, and experiences within the topic areas covered, allowed us to be more intentional about the inclusion of relevant personal thoughts and experiences. "[A]utoethnography involves the researcher's shifting back and forth between the social and cultural and their inner experience" (van den Hoonaard & van den Scott 2022: 215). It changes the research project so that in addition to studying someone else, the researcher's own observations, actions, and feelings are analyzed and included (van den Hoonaard & van den Scott 2022: 215). This also allowed us to work through some of the difficult emotional labor that arose in this research.

In exploring others' experiences of marginalization, structural violence, systemic inequality, and climate disruption, we often felt a sense of profound overwhelm and grief. Researchers are often expected to be "objective" and "unemotional", but the emerging thought is that this is not beneficial, or perhaps even possible, especially when witnessing violence and loss (Gillespie & Lopez 2019). Such emotional labor may also bring to attention our own privilege as "researcher" and "witness"' as opposed to those who are "researched" and "witnessed" (Gillespie & Lopez 2019: 7). We felt this deeply in our own research as we thought through the privilege we have currently as white, middle-class, educated professors living in the Global North exploring what food in a just world means, in addition to our species privilege as humans compared to the other animal subjects we tried to bring to the forefront in this work.

An intersectional framework exploring social justice necessitates that we expand the critical lenses to explore how "oppression in all forms intersect, and that actors in the more-than-human world are subjects of oppression and frequently agents of social change" (Collins & Bilge 2020: 239). This project started from a perspective of interdependence, the view that humans are part of an intricate network of relationships with each other, with other animals, and with the environment, and that harm to one part of the network inevitably affects all other parts.

Such sharing of emotional labor is also important to us from an emancipatory standpoint, as being active in the research process and accepting the rigor and substantiation of works that are in addition to those more traditionally used in academe, is an important part, we believe, of the decolonization process. As Tilley (2016) points out, "[c]ritical frameworks create space to question dominant, historical Eurocentric practices embedded in the institutions within which many students learn to be researchers and receive permission and resources to conduct research" (p. 11). We set about in this project to challenge those dominant assumptions about who and what should be considered in questions related to building a just food system by expanding the emancipatory lens to prioritize marginalized voices, viewpoints, and experiences throughout the book.

During her sabbatical in the winter of 2021, Tracey began exploring research on the treatment of nonhuman animals within industrialized food systems and the job-related outcomes for slaughterhouse workers, especially during the COVID-19 pandemic. Meanwhile, Terry had been working closely with local Indigenous knowledge keepers as a participant in the movement to decolonize settler mindsets and institutional frameworks. This book project, and the qualitative data collection used within it, started from the standpoint that Indigenous knowledge systems are critical to questions of food justice and resilience in that the lenses framing humans in relationship to other humans, nonhuman animals, ecosystems, and the land, fundamentally challenge the logic of the capitalist system and the inevitable harms of a profit-motivated system.

The inclusion of Indigenous knowledge systems was essential to this research project as Mi'kmaw Elders and community members, and Indigenous Peoples globally, are leaders in protecting the integrity of ecosystems and ensuring that future generations will be able to live healthfully and compassionately on this planet. We similarly come from a place of understanding the important relationship of ecosystems to humans and nonhuman animals as an interdependent system where both harms and benefits in one part of the system affect all other parts, and that the goal should be sustainability and compassion for all.

One aspect of this research has been to draw on these important understandings to help others (i.e., non-Indigenous) understand the importance of Mi'kmaq and other Indigenous knowledge systems in creating more just food systems (Battiste 2013; Kimmerer 2013). If the ideas presented in this book lead to more sustainable, compassionate, and healthier approaches to food advocacy, and to healthy food options becoming more affordable and easily attainable, it would be of benefit to all communities, not only Mi'kmaq and other Indigenous communities.

The research for this book looks to the experiences of advocates and researchers of Black, Indigenous, and other equity deserving groups, Mi'kmaw Elders and Knowledge Keepers, farm sanctuary workers, regenerative farmers, producers, policymakers, consumer-citizens, students, and youth in general. Each participant helped us to better understand the issues with the current food system and the ways in which people are challenging it and creating more equitable and compassionate alternatives.

At the end of the research process, and reflecting on all that we have learned, we hope to have provided a broad picture of the many reasons the current food system cannot continue to be governed by an economic model guided by a logic of profit that often treats humans, other animals, and nature as commodities; not if what we seek is to create a more just, compassionate, and sustainable world for all. For a more detailed explanation of the methodology, please see the Methods section at the end of the book.

Introduction

We would like to begin by acknowledging that we are writing this book from Mi'kma'ki[1] and that, as a result of unlawful trespass and dispossession, settlers are here on the territory of the Mi'kmaq. The land and territories were never ceded by treaty nor have been purchased by the British Sovereign. Mi'kmaq Chiefs in the treaty reserved their ancestral territory and resources for their clans and families, and for "their heirs, their heirs of their heirs forever" (Treaty 1726, renewed in 1752). Mi'kmaq Chiefs also authorized peaceful British settlements in Mi'kma'ki as trading posts. However, the settlers gradually dispossessed the Mi'kmaq. The Peace and Friendship treaties lay the foundation for relations with settlers, and guarantee the inherent powers and territories for the Mi'kmaq Nation. Reconciliation then requires us to respect each other, work together for the benefit of all, and to take care of the earth. This opening acknowledgment is not only an important reminder of our place as authors, and our history, but it also relates to so many aspects of this book focused on creating a more compassionate and just food system.

We begin from the standpoint that Indigenous knowledge systems are critical not only to questions of social justice related to food but as foundational lenses for interdisciplinary discussions and collaborations on the broader questions of climate change and global citizenship. These lenses frame humans in healthful and compassionate relations with other humans, nonhuman animals, and the land in ways that fundamentally challenge the harmful and often violent features of our global economic system driven by profit and

---

[1] Mi'kma'ki includes seven ancestral districts that include all of what settlers called Nova Scotia and Prince Edward Island, as well as parts of New Brunswick, Quebec, and Newfoundland in Canada and part of Maine in the United States.

overconsumption. The inclusion of Indigenous knowledge systems was also important to us as we work in solidarity with Mi'kmaw and Indigenous Peoples globally who are leaders in protecting the integrity of ecosystems and ensuring future generations will be able to live healthfully and compassionately on our planet.

It is important to note that Indigenous communities, not only on Turtle Island (North America) but around the world, are disproportionately affected by industrial production systems such as resource extraction and food production that have violated Indigenous rights and expropriated Indigenous lands. In addition, one of the clear outcomes of the industrialized production of other animals as food is climate change which, once again, disproportionately impacts Indigenous communities (LaFortune 2020).

Since food, clean water, and fresh air are all basic necessities of life, all communities have a vested interest in seeing the climate crisis resolved. With this in mind, it was important to us to frame Indigenous knowledge systems and viewpoints as essential in constructing sustainable and compassionate ways forward. This is also essential within the Canadian context as there has been a "Call to Action" in response to the history of colonialism (Truth and Reconciliation Commission of Canada 2015), made more blatantly visible in the past few years through the gruesome discoveries of mass graves of Indigenous children at residential schools. This brutal consequence of systemic abuse and neglect illustrates the urgent need to decolonize policies, frameworks, belief systems, research, and institutions in order to foster true and lasting reconciliation.

The book that you are reading has evolved over time. At first, we thought it would be an overview of the many ways that the Animal-Industrial Complex (A-IC) – the large-scale, industrial system of production, distribution, and consumption of animals as "food" within the broader global capitalist system – harms humans, other animals, and the ecology of the planet (Noske 1997; Twine 2012). But the research was telling us that while there are many excellent resources that explore the systemic issues and impacts connected with the industrial production of animals as food in general, there was little out there that weaves nonhuman animals into the broader narratives of social and climate justice in a more holistic way (i.e., not just as an "add on" or for a readership primarily interested in nonhuman animals).

Many authors before us, several of whom we had an opportunity to interview for this book, offer comprehensive studies on the harms of the A-IC and the intensive system of producing other animals as food. We are cognizant of, and fully appreciate, the influence of these other studies in laying the groundwork for this book and their voices are weaved throughout.

We chose the title *Food in a Just World* rather than "Food Justice" intentionally to highlight the reality that building a genuinely "just" food system is inseparable from building a world in which broader social justice issues related to racism, classism, homophobia, transphobia, ableism/disability, sexism, and speciesism are all tackled together. While it is obvious to us that the current dominant food system directly and unapologetically makes nonhuman animals into commodities, there are few voices out there that demonstrate how this reality is interwoven with and inseparable from the commodification of people and land.

We would argue that the food justice issue is connected with all other social justice struggles and, directly, to biodiversity destruction, mass species extinction and climate change. These realizations created a two-fold issue for us in writing this book. Firstly, we feel we cannot get to the heart (literally) of the matter of food without a very broad scope, but we also acknowledge that we cannot in one book adequately address all social justice issues or the literature of all the disciplines that tackle the food question. So, our caveat is that we don't pretend to!

We apologize in advance for the discomfort that this may create for some readers, particularly the academic ones. What we intend to do, rather, is to provide a guide to some key threads that we have identified as intersecting amongst the various literatures and disciplines, pointing where possible to helpful resources along the way. In addition to that, we will highlight some inspiring examples of the drivers of change.

This book is therefore, by nature, interdisciplinary and intersectional with a focus thematically on rethinking democracy – in the largest sense of the word – through the lenses of social and ecological justice and as a process which must include complete transparency and accountability. It tackles several of the intersections of oppression related to the industrial food system and points to ways in which seemingly divergent issues are related. Uniquely, we believe, it examines some of the issues of the heavily industrialized and intensive food system from multiple and connected perspectives: citizen-consumers, workers, nonhuman animals, and the environment.

Fueled by firsthand interviews with knowledgeable experts and advocates who work directly on food and environmental issues, we further draw upon, and attempt to synthesize, work from a variety of other sources: scholarly work, policy papers, news articles, and popular information sources such as documentaries. We also hope that this work will inspire and demonstrate that, despite the precarious and unknown future of life on this planet, not only is positive change possible, but as the interview participants point out, in many communities it is already happening.

Our book assumes that there are three key parts to how change happens. One is through the activism, advocacy, and emotional and practical labor of concerned citizens from all walks of life – classes, races, genders, and cultures. These advocates are trying to make transparent, and end as many of the harms, particularly the worst harms, of the system as quickly as possible. Another significant and often forgotten or ignored way that change happens is through the modeling of new ways of being by individuals, communities, and organizations demonstrating practical alternatives. The third aspect of shifting oppressive systems is large-scale institutional processes and legal changes (i.e., changes at the structural level). This final category of change often frustrates people from all places on the political spectrum because it is the one in which we can often feel the most impotent.

In the book we give attention to all three of these aspects of change, again with the proviso that we do not claim to adequately explore all the facets or implications of the global structures of governance or economy, or all the incredible work for change in food systems happening around the world. The food system, treatment of workers in producing all types of food, but especially those that relate to nonhuman animals, and climate disruption are all increasingly receiving attention; an attention that has intensified since the beginning of the COVID-19 pandemic. We believe it is important to keep talking about these intersecting issues while amplifying the success stories, and that takes a broad lens.

Dr. A. Breeze Harper, Critical Race Feminist and Food Studies Scholar, said in her interview with us that "dystopian stories make us feel helpless and hopeless. We need to demonstrate a reimagining that a 'utopian regenerative future' is possible. This will bring much-needed hope and resilience to the discussions of food systems, and environmental care" (Interview with Breeze Harper 2022). And, as we heard from many interview participants, "paradigm shifts" also need to move the discussion so that marginalized voices and wisdom are front and center.

We also need to model compassionate ways forward, by telling the stories of those most disadvantaged in this current food system – humans and nonhumans alike – while offering up potential solutions that are currently initiating, or could soon initiate, positive social change. When we asked the filmmaker of *Speciesism: The Movie*, Mark DeVries, what was something he now knows that he wished he'd known when starting his work/research, he responded that, first, reforming the food system is complicated, stating:

> For every change that we make, we should expect unanticipated consequences, which may include problems that must be solved in turn. We should always remain intellectually humble, and open to new information that provides reasons to update our views. Second, change is happening, even if it may not appear to be happening fast enough. Taking a long-term view can allow us to feel calm and remain engaged, rather than feeling overwhelmed and getting burnt out. (Interview with Mark DeVries 2022)

We hope that this book humbly considers some of the ways the food system is being challenged, how it has been positively transformed, and ways to further transition. To do this we present material from a variety of sources and places. The interview participants include an Indigenous Elder; racialized and other equity deserving peoples; advocates; researchers; allies; academics from a variety of fields related to food, ethics, and justice; farmers and producers of organic and sustainable food systems; animal rescue and sanctuary workers; students; politicians; and educators.

All this is to say that we have talked to inspirational thinkers and doers who are helping to re-shape the food system to one that takes the well-being of nonhuman animals, workers, consumers, and the environment into consideration. These voices have helped to shape this book, and we are grateful for the time, energy, and inspiration these participants have shared with us. From a personal perspective, we also want to acknowledge what a privilege it has been to talk to, and learn from, such inspirational people in the creation of this book.

Throughout the book we attempt to fully consider the lives of nonhuman animals under the industrialized food production system. One of the ways we do so is to bring attention to their commodification, and the fact that this is neither "normal" nor inevitable. One way of drawing attention to this is to consider the ways in which we talk and write about nonhuman animals, because the language we use helps us create, shape, and share cultural ideals and values. Throughout this book, we utilize the terms "nonhuman animals" or "other animals" to help signify that we too, as humans, are animals.

As David Nibert so eloquently argues in his work, in addition to reminding us that we are also animals, such phrases offer us "a way to challenge the entrenched patriarchal, dualistic way of perceiving the world as involving 'man' (purportedly rational and 'civilized') on the one hand and women and other animals (inferior) on the other" (2017: xviii). He also reminds us that even if the intent in using this terminology is to overcome speciesist language and understandings, the phrase 'nonhuman animals' is still imperfect because it still prioritizes humans. It also lumps together an enormous grouping of beings – some quite similar to us (such as chimpanzees) – and the term does not recognize the differences between species in both personalities and in power relations with humans (for instance, a dog and a chicken) (Gruen 2015: 35–36).

In addition to land animals, we also look at the utilization of fish for food. Both Jonathan Balcombe (2016) and Taichi Inoue (2017) urge us to consider using the plural "fishes," rather than the common usage of the singular, when describing entire species. As Balcombe states, "I have come to favor the plural 'fishes,' in recognition of the fact that these animals are individuals with personalities and relationships" (2016: 6). Both authors believe that it also helps us to recognize that fishes and other sea animals matter, beyond their utility to humans and their ongoing commodification.

Throughout the book we try to draw the reader's attention to mainstream language that devalues other animals, or frames them only within the conceptualization of them as "food," "tools," "property," etc. We do this by italicizing commonly used terms like *meatpacking, chicken processing, farmed animals, livestock,* or particular products made from their bodies or output, such as *meat, veal, cheese, omelet,* or *hamburger*. We rarely use the term "farmed animals" and instead use the term "industrial production of animals as food." The hope is to avoid objectifying and defining other animals by the nature of their oppression and recognize them instead for the amazing and diverse individuals they are (Nibert 2017: xviii). We recognize that this can make reading the work a bit more clunky or cumbersome, but we believe it is important to demonstrate the problem with the usage of such words as we try to build a more respectful and compassionate relationship between human and nonhuman animals (Gillespie 2018: 8). We also mainly use the terms "slaughter" and "slaughterhouse" to describe the process of turning live animals into *meat* or *food*. Such terminology may not be common in industry, and other mainstream narratives, but it does provide "clarity in describing what transpires within the confines of those walls" (Fitzgerald 2015: 35).

Making visible that which is often invisible is also an important feature of intersectional and emancipatory research. As sociology professors Patricia Hill Collins and Sirma Bilge explain, "complexity is not something that one achieves by using intersectionality as an analytic tool, but rather something that deepens intersectional analysis" (Collins & Bilge 2020: 34).

We also want to share here a little bit about who we are as authors and how we came to create this book. While neither of us believes in the possibility of conducting unbiased fieldwork and research, we do believe in the power of putting all our cards on the table, so that the reader has a clear picture of who we are, and how we have come to think the way we do. While we each have our own stories to tell, we will start with an acknowledgment of the similarities in our life histories that allow us to easily work on this project, and many others, with shared insights and conviviality, and, when appropriate, good humor.

We come from different academic backgrounds, but we both share a commitment to interdisciplinary and intersectional approaches. We both also share a commitment to emancipatory ideals in our teaching, research, and community service. And while this gets implemented in diverse types of courses, research, and service, it is the foundation that allows us to build a bridge that connects the various pieces of our work and advocacy.

Both of us grew up in lower-middle/working class "meat and potato" families without any understanding of vegetarian or plant-based diets or vegan lifestyles. Terry remembers, when her mother became a single parent and the family lived on social assistance temporarily, the family had a "budget box" with coin slots where a certain number of coins went to "meat" and some to "vegetables" and quite a bit less went to "treats." Tracey recalls her parents having a little box that contained envelopes where the weekly allowance for each necessity was carefully measured out in order to stay on budget for the week. This adds another layer to our understanding of the necessity of not assuming that everyone is starting from the same place or will end with the same behavioral shifts.

Although we grew up in vastly different geographical locales, we sometimes wonder at the similarities between our rural upbringings. For instance, we were both first-generation university students, we both grew up with a huge appreciation and respect for other animals and the natural world, and we lived in close-knit, rural communities where people helped each other out. Terry spent her very early childhood in three Arctic hamlets in what is now known as Nunavut and was deeply impacted by Inuit culture and Inuit practices with regard to the land and other animals. But having a deep-seated appreciation for nonhuman animals did not mean that we were always able to fully see what their lives entailed. For instance, shortly after beginning this book project we were having a conversation about *livestock auctions* in both Canada and Wales and realized that, as children, both of us had frequented these so-called *cattle sales* as spectators. As children we were admittedly naïve to what we were witnessing, and both of us loved to attend in order to visit with the animals and muck around in the facilities.

Interestingly, and connected to this current project, there are several more similarities on a personal level that connect us together and to this project. We are both mothers and feel deep sadness at what is happening to the planet and a sense of urgency to leave things in some semblance of "good shape" for our children, and their children, and their children's children, and so on. We both eat only plant-based foods and strive for a vegan lifestyle, but our children are vegetarian, and our partners are omnivores. As we are both the primary cooks in our households, the effect is that each of our families eats a primarily plant-based diet.

We have seen firsthand the positive health effects of this diet on our partners, in terms of lessening certain hereditary diseases like early onset heart disease, and lifestyle diseases such as high cholesterol and diabetes. With family and friend configurations, including children of mixed race/ethnicity, we also have firsthand experience with navigating dietary differences and preferences, and have become self-taught "experts" at veganizing meals. Furthermore, with advice from friends, we love experimenting with veganizing recipes from diverse cultures, an important and creative challenge highlighted in the interview with Mi'kmaw scholar Margaret Robinson (Interview with Margaret Robinson 2022).

We fully acknowledge that we don't always make the grade in our food/consumer choices, and we certainly do not feel "holier" because of our choices! We acknowledge that every individual who actually has a choice decides on certain priorities that differ and, at times, may not make sense, to another person with similar circumstances or values. We discuss this further in the book.

Almost all of the interview participants pointed to the fact that taking social justice into account with every dollar or other currency we spend is *very* difficult, and sometimes impossible, in this current economic system. And, as we argue in the book, "choice" is often a feature of privilege. But we do believe that navigating these differences in our family members' diets gives us a unique understanding of some of the issues people may face in thinking about, or transitioning to, more plant-based foods and ecologically friendly diets and lifestyles, and how we can be helpful allies for those trying to incorporate such changes into their lives.

One of the struggles we faced in the layout of this book was trying to present concrete solutions for the vast array of issues facing humans, other animals and the environment within the parameters of the current animal-intensive food system. We recognized it is essential that all three pieces come together and are seen as equally important. It was also a struggle to reconcile how to present partial or incremental change for the betterment of nonhuman animals, without legitimating their subordination or continued (ab)use. We do, however, clearly recognize the fine line that must be walked in figuring out how to support incremental change, especially as it relates to animal *welfare* measures, without them becoming a rationale for continuing with the status quo.

We have seen that one way to bring together people from various belief systems and vantage points is to focus on the concept of transparency within the food system. In her interview with us, sociologist and green criminologist Amy Fitzgerald said, "One thing I tell my students is to gather information about what they consume. And I always emphasize to them that we should all be troubled, even if they do not care about animals, by the lack of transparency. This is an issue that should unite people" (Interview with Amy Fitzgerald 2022).

We recognize that such transparency is not easy as it is often difficult to ascertain what actually happens in the production of food. But this difficulty elevates the importance of transparency because it spurs us to ask: Why is it so difficult to find out how food is produced?

Within this book we will address the roots of this lack of transparency and will advocate for the kinds of changes other researchers and advocates propose to challenge this system, both from a gradual incremental perspective and from a more system-overhaul, emancipatory approach. We attempt to do this while maintaining a firm belief in building bridges, providing education, and acting and reacting compassionately. In all our work – whether research, community service, or teaching – our commitment is to try to reach people where they are at. We strongly believe that most people do not respond well to being shamed, and that real and lasting change requires time, energy, community building, and active listening to disparate viewpoints. We very much hope that this will come through as you begin reading.

In chapter 1, drawing on Indigenous and other knowledge systems and perspectives, we explore what a more genuine and deep form of democracy and democratic accountability could look like and what it could do to transform the way we think about food. In this exploration we make use of the concept of "structural violence" (i.e., the violence, both direct and indirect, that exists in systems), and we situate this discussion in the context of the current global economic system of capitalism which has its own internal logic.

Contextualizing the discussion in the framework of climate change and biodiversity destruction, chapter 2 looks at the 'Animal-Industrial Complex' (A-IC) as one significant feature of the global economic system. It also sketches out the impacts of this system for nature/land and for nonhuman and human animals. With food in mind, we examine colonialism and linear notions of "development," arguing for a shift away from "dominion" over land and other animals towards a culture recognizing and protecting the global "commons," neutrally defined by the *Oxford English Dictionary* as "land or resources belonging to or affecting the whole of a community."

Chapter 3 looks at work within industrialized production of other animals as *food*, for both humans and nonhumans. Using a case study approach, this chapter focuses on some of the systemic issues with this type of work, how those issues were exacerbated during the initial COVID-19 pandemic and how they may threaten to play out in future outbreaks and health crises. It concludes by examining resistance and the possibility of what a just transition away from industrialized production of animals as *food* might involve and what it could look like.

Chapter 4 examines the current animal-intensive *food* system from the frameworks of a colonial-capitalist diet, and what that means for consumer-citizens. Here we explore food as a basic human right. The rise in industrialized production has not led to everyone being well fed, and a host of other concerns are also touched upon, such as food scarcity, overabundance of cheap processed food, inequitable food distribution, and food safety. We briefly reflect on some of the recent discussions of zoonosis, when animal diseases cross over the species boundaries to humans, and the potential consequences.

Chapter 5 examines what a food system that commodifies other animals on an industrial scale looks like more specifically for nonhuman animals. It highlights the connections between what are known as *Standard Industry Practices* in countries such as Canada, the United States, and in the European Union. This means that, while individual workers and consumers may not set out to be intentionally cruel to other animals, the profit-driven system makes certain decisions and practices – practices that often do not reflect a concern for the well-being of nonhuman animals – more convenient and more affordable. We end this chapter by focusing on the well-being of nonhuman animals as part of a larger critique of what needs to change to facilitate democracy in practice.

The final chapter explores the possible features of a compassionate food system, highlighting the Mi'kmaw concept and practice/embodiment of "Eptuaptmumk" or "Two-Eyed Seeing" and advancing the notion of "radical democracy." In the vein of ensuring transparency, and as a prerequisite to social and ecological justice, we advocate for a *precautionary* approach to economic development where the goal is to minimize and, where possible, to eliminate, harms to the earth and to all animals, including us humans.

This final chapter also considers some of the ideas proposed by interview participants as they reflected on the need for a paradigm shift in the transition to more democratically decided and locally based food strategies. The conclusion folds these various themes together, highlighting the challenges and opportunities ahead in the process of democratizing the food system and pointing to areas for future work and research.

1 Food Justice Needs a Just World

Confronting Structural Violence against Land, Humans, and Nonhuman Animals

> *I learned very early that the violation of ecological limits is the beginning of ecological injustice. And this is what has shaped my understanding of climate change.*
> – Vandana Shiva

In this book we are not simply interested in examining food as an isolated phenomenon to be explored in relation to other social determinants of health and well-being. We are proposing that a socially just, ecologically responsible and transparent approach to food systems is a fundamental, if not a *necessary* prerequisite, to building more democratic, resilient, and compassionate communities where well-being for all, including other species, is a guiding principle. We will explore below why food security depends upon food "justice" which in turn depends upon genuinely democratic, from the ground up, political and economic systems that take a precautionary approach with regard to nature and all living things.

We live in a world that produces enough food to feed everyone and yet the United Nations Food and Agricultural Organization (FAO) reported that there were 828 million people affected by hunger in 2021; an increase of 46 million people over the previous year and 150 million people since 2019. The report further estimates that 3.1 billion people do not have access to a healthy diet. At the same time, the global waste of food – post-harvest and prior to consumption – is around 14 percent. Meanwhile, according to United Nations Environmental Program's (UNEP) *Food Waste Index Report*, another 17 percent of food is wasted in retail and by consumers. It is estimated that, with the food we waste alone, we could feed 1.26 billion hungry people every year (UNEP 2021).

In its ground-breaking report of 2013, the FAO put the global volume of food wastage at 1.6 billion tonnes, with the estimated carbon footprint of this wastage at 3.3 billion tonnes of greenhouse gases (GHGs) released into the atmosphere. Astoundingly, we use almost 30 percent of the world's agricultural land to produce wasted food. In addition, "Agriculture is responsible for a majority of threats to at-risk plant and animal species tracked by the International Union for Conservation of Nature" (FAO 2013). This picture has not shifted significantly in recent years as it is estimated that food wastage accounts for 8–10 percent of global GHG emissions (FAO 2019).

At the World Food Summit in 1996, it was declared that, "Food security exists when all people, at all times, have physical and economic access to sufficient, safe and nutritious food that meets their dietary needs and food preferences for an active and healthy life" (quoted in FAO 2006). The definition does not say "some people, sometimes," nor does it say "mostly white people," or "for part of their lives"; it clearly states *"all people, at all times"* (our emphasis).

Drawing on statistical data and reports from reputable global food institutions, in addition to interview testimonies, we argue below that the current food system simply does not work for most of humanity and is at the same time systematically destroying biodiversity and animal life while being a disproportionate contributor to climate change. Further, as we shall see, the statistical correlations of class and race to food insecurity require that we explore issues of poverty and social justice more broadly speaking, as we attempt to address how to improve food systems.

A stark glimpse of this picture is evident in the lives of small farmers, who, analysts suggest, are responsible for 30 percent of the food supply globally. Hannah Ritchie from *Our World in Data* notes that, "Most (84%) of the world's 570 million farms are smallholdings; that is, farms less than two hectares in size" (Ritchie 2021). These largely rural populations face the most acute poverty and hunger. We will explore below the structural conditions that make this paradox so. In their recent report, the FAO and the World Food Program (WFP) were warning that "acute food insecurity" globally would continue to escalate with the situation likely worsening in 19 countries, known as "hunger hotspots." These include those countries projected to face starvation: Afghanistan, Ethiopia, Nigeria, South Sudan, Somalia, and Yemen (FAO & WFP 2022).

Global organizations are increasingly tying their escalating numbers of those facing food insecurity to climate change. In 2017, the WFP was reporting that over 20 million people in the African countries of Ethiopia, Malawi, Zimbabwe, and Kenya experienced acute food insecurity as a direct consequence of climate change. The WFP, which has a "special relationship" with the UN Food and Agricultural Association, notes that climate disruptions such as La Niña since late 2020 are causing massive crop and *livestock* losses, particularly in East and West Africa, Central Asia, and Central America and the Caribbean (FAO 2022).

In most of the world, food security is dependent upon access to fertile agricultural land. Consequently, when people are forced to flee their lands they inevitably face increasing food insecurity. In 2021, the White House reported that, along with conflict, climate disruptions are one of the top two drivers of forced displacement globally, together causing nearly 30 million people to flee their homes and lands annually (The White House 2021).

The loss of glaciers, a process driven by climate change, is just one element of this picture of food insecurity globally. The melting of glaciers is often perceived as a distant threat, and the links of this to food systems are poorly understood by most people. Mount Kenya provides a stark example of this. As Africa's second-largest mountain after Kilimanjaro, its various peaks are projected to lose most, if not all, of their glaciers by 2030, making it the planet's first mountain range to lose its glaciers to climate change. The International Livestock Research Institute (ILRI) notes: "The Mount Kenya ecosystem is an irreplaceable biodiversity hotspot that provides water for over 2 million people, and the surrounding landscape has long been one of eastern Africa's most productive agricultural areas" (Carleton 2022).

According to UNEP, the broader African continent is responsible for a paltry 2–3 percent of global greenhouse gas emissions, but it is the most vulnerable region globally in terms of climate change and food insecurity. The connection of this vulnerability to global inequality and poverty is a pattern repeated around the world, making Global South populations (particularly around the equator and island communities) and Northern territories, in places such as Canada, the most at risk.

It seems appropriate to state that the people impacted by these processes are victims of "structural violence," not to be polemical, but to allow us to identify the harmful elements of "systems" (i.e., political and economic structures) that are actively contributing to these processes. Structural violence is a term that grew out of Marxism and then was taken up by Norwegian sociologist Johan Galtung, principal founder of the discipline of peace and conflict studies. Galtung showed how "violence" can take many forms and can be expressed in indirect ways through institutional, bureaucratic, and cultural practices (Galtung 1969).

The institutional infrastructure required by the Nazis and used during European colonialism are two poignant examples. This is important because if we can't recognize how political and economic institutions and processes allow, enable, or in some cases directly cause violence, we may end up with any number of outcomes including genocide. And it is important to understand why average people can be working in service of violence without necessarily always being aware of that fact.

Structural violence is not only an issue for people in the Global South. In many countries of the relatively rich Global North, systemic racism is prevalent and directly relates to food insecurity. In Canada, for example, household food insecurity is highest amongst Indigenous and other racialized groups (PROOF 2022). Not surprisingly, class is also a huge factor in food security. When highlighting who are most at risk of household food insecurity generally, the policy-focused interdisciplinary research group PROOF points to those with "inadequate, insecure incomes and limited, if any, financial assets, or access to credit" (PROOF 2022). This disproportionate pattern of food insecurity is also experienced by racialized and economically disadvantaged groups across the United States (Haider & Roque 2021) and the UK (PROOF 2022).

The reality of the racialized and inequitable dimensions of food security, in combination with the lack of transparency and appropriate responsibility being taken around impacts and responses to climate change, makes clear why food is a social and ecological justice issue. In other words, "food security" is about a lot more than simply feeding ourselves. It is about how we produce the food we eat, what and whom we eat, and how this food is distributed. In all of these questions, we argue below, a transparent and "precautionary" approach, grounded in more localized forms of decision-making, will be the most effective in terms of ensuring that more equitable, racially just and ecologically sound practices are pursued. With this in mind, we begin with a discussion of the land.

Climate Change and our Relationship to the Land

As we edited the drafts of the first chapters for this book, Cape Breton Island, or Unama'ki as it is known by the Indigenous Peoples of this land, was recovering from a battering by Hurricane Fiona. This was arguably the most serious storm on record to hit Canada's Atlantic shores. Climate scientists and meteorologists pointed out that the timing, power and devastation of this storm were unprecedented. These impacts are directly related to the warming waters of the North Atlantic and foliage not normally on trees to this degree when storms of this nature hit.

Climate scientists have stated that we should expect these kinds of disruptions, and possibly even more calamitous versions of them, more frequently in the coming years (NASA 2023). So, the starting point for any discussion about food is that the land we will need to depend upon to ensure future food security is itself undergoing multiple transitions, and in some places abrupt change. This is occurring all over the world and some of this change is not neatly predictable (Jamail 2020; Rushkoff 2022).

We can depend upon the fact that relations with the land, other humans, and all the nonhuman animals with whom we share this planet are to some degree always changing. Climate disruption, however, brings this reality more sharply into focus. As military and international organizations around the world warn us, climate change threatens to be the chief "security" problem in the future. US Defense Secretary Lloyd Austin has called it an "existential threat." How do we better prepare for this change so that we not only can continue to produce healthy food, but ensure that we have the mutual supports in place to navigate the coming changes as compassionately and equitably as possible?

We know that sea levels are rising, that glaciers are melting, and that climate change and human production and consumption patterns are causing the destruction of nonhuman animal habitats around the world, with plant and animal species being wiped out by the thousands every year. When peer-reviewed research from scientists, claiming to be "conservative" in their analyses, tells us that species population declines and destruction can be described as a "biological annihilation" and "a frightening assault on the foundations of human civilization," it behooves us to pay attention (Ceballos et al. 2017). Besides the obvious horror of systematically destroying the lives of other animals, we have a selfish reason to listen because humans depend upon the "ecosystem services" that other animals maintain, particularly when it comes to feeding ourselves (Ceballos et al. 2017).

The US-based National Wildlife Federation (NWF) notes: "An ecosystem service is any positive benefit that wildlife or ecosystems provide to people. The benefits can be direct or indirect – small or large" (National Wildlife Federation 2023). The NWF identifies four main areas of "services" provided by wildlife and ecosystems. These include "provisioning services," such as providing the fruits and vegetables we eat as food; "regulation services," for example the work plants do to clean air, filter water, create bacteria to decompose waste, bees to pollinate flowers, apex and other predators that keep smaller populations within an ecosystem in balance, and tree roots to hold soil in place and prevent erosion. There are also "cultural services" related to things like recreation, physical activity, mental health, art and music, cultural heritage, and spirituality. And, finally, they involve what the NWF calls underlying "supporting services," which are seen in "natural processes, such as photosynthesis, nutrient cycling, the creation of soils, and the water cycle. These processes allow the earth to sustain basic life forms, let alone whole ecosystems and people. Without supporting services, provisional, regulating, and cultural services wouldn't exist" (National Wildlife Federation 2023).

These ecosystem services have been disrupted in the past. For instance, five "mass extinction" periods have occurred during our fossil history, but the current one is the first mass extinction to be caused by humans. As noted by the World Wildlife Fund:

> The sixth mass extinction is driven by human activity, primarily (though not limited to) the unsustainable use of land, water and energy use, and climate change. Currently, 40% of all land has been converted for food production. Agriculture is also responsible for 90% of global deforestation and accounts for 70% of the planet's freshwater use, devastating the species that inhabit those places by significantly altering their habitats. It's evident that where and how food is produced is one of the biggest human-caused threats to species extinction and our ecosystems. (World Wildlife Fund 2023)

Earth Democracy and the Democracy of Species

In reference to trees, children's book author Clyde Watson asks, "What is it like to bend and sway but not be able to run away?" (Watson & Morse 1976). In this question she captures the way many of us feel in the face of climate change and the very real threat it poses to human survival on this planet. The question also speaks to the "aliveness" of the plants and individual nonhuman animals around us, many of whom are eaten as *food*. This aliveness is something that is self-evident to Indigenous Peoples.

As Mi'kmaw Elder Albert Marshall points out, before we can think about food and how to create more democratic and compassionate food systems, we must first understand the interdependent relationship with the land and with all the living creatures and plants with whom we share this planet (Ongoing collaborative work and research with Albert Marshall on "decolonization" and conversations recorded in spring 2022). In a similar vein, Robin Wall Kimmerer, Professor of Environment and Forest Biology, and enrolled member of the Citizen Potawatomi Nation, notes: "The traditional ecological knowledge of the indigenous harvesters is rich in prescriptions for sustainability. They are found in Native Science and philosophy, in lifeways and practices, but most of all in stories, the ones that are told to help restore balance, to locate ourselves once again in the circle" (Kimmerer 2013). In Kimmerer's recent work, she states: "In the indigenous view, humans are viewed as somewhat lesser beings in the democracy of species" (Kimmerer 2021: 86).

In our approach to compassion, and to compassionate food systems in particular, we are inspired by and feel a deep resonance with Kimmerer's notion of a "democracy of species," and by Indian scholar, physicist and environmental activist Vandana Shiva's concept of "earth democracy" (Shiva 2005). In *The Democracy of Species*, Kimmerer explains, "The story of our relationship to the earth is written more truthfully on the land than on the page. It lasts there. The land remembers what we said and what we did. Stories are among our most potent tools for restoring the land as well as our relationship to the land. We need to unearth the old stories that live in a place and begin to create new ones, for we are story makers, not just story tellers. All stories are connected, new ones woven from the threads of the old" (Kimmerer 2021: 76).

The democracy of species, from this perspective, could provide an orientation for the political and economic world that understands relationships to the land and other animals in this way, as story makers. It can also guide us with regard to respecting the plant life that supports healthy ecosystems and the nonhuman animals who provide both the ecosystem services as parts of "essential species" and all other individual animals who have relationships, complex emotions, and who strive to live the best lives they can (Bekoff 2007; Balcombe 2010).

From a slightly different, although in our view entirely compatible, angle, Navdanya International in India works at the grassroots with local communities to build "earth democracy" as conceived by Vandana Shiva. Navdanya

> works to promote a shift towards a circular model where the economy is held in no greater esteem than the environment and the resources it provides and the people, for whom justice, fairness, and the right to resources is of equal importance. It's a new vision for a Planetary Citizenship, an alternative worldview, rooted in caring and compassion for the Earth and Society, in which humans are embedded in the Earth Family, where ecological responsibility and economic justice, based on the Law of Return, are central to creating a liveable future for humanity. (Navdanya International 2023)

As Shiva explains:

> Earth Democracy connects the particular to the universal, the diverse to the common, and the local to the global. It refers to what in India we refer to as "vasudhaiva kutumbkam" (the earth family) – the community of all beings supported by the earth. Native American and Indigenous cultures worldwide have understood and experienced life as a continuum between human and nonhuman species and between present, past, and future generations. (Shiva 2005: 1)

Navdanya's projects have allowed for the collaboration of thousands of Indian farmers and have supported them in the transition from industrial monoculture – systems based on pesticides and genetically modified organisms (GMOs) – to biodiversity-intensive agroecology (Navdanya International 2023). Terry has conducted research at Navdanya and developed a version of her course on *Global Citizenship and the Compassionate Society* to take place there, enabling students to live, study, and engage in work in India for a short time.

In addition to Navdanya, it is important to note the personal achievements of Shiva as they indicate the global recognition, resonance with, and possibilities of the transformations she is working to promote both as a scholar and an activist. For instance, in recognition of her activism on behalf of women and ecology, Shiva received the esteemed *Right Livelihood Award* in 1993 (also known as the alternative Nobel Prize). She has also been identified as one of the globe's most important activists by media giants such as *Time*, *The Guardian*, *Forbes*, and *Asia Week*.

While it is beyond the scope of this book to explore all the global manifestations of this kind of thinking and work, it is clear that the spirit of interconnection and interdependence evident in the ideas of *Earth Democracy* and the *Democracy of Species* is echoed in the language and practices of cultures around the world; for example, in the concept of "Ubuntu" in southern Africa, captured by the late Archbishop Desmond Tutu in the phrase "I am because we are"; and in Vietnamese spiritual leader Thich Nhat Han's parallel idea of "interbeing" (2001). These are two concepts that have been explored in many other published works. The idea of interdependence for some Indigenous Peoples is symbolically captured in the expression "msit no'kmaq" translated as "all my relations," an idea central to First Nations, Inuit, and Métis Peoples in Canada in which nonhuman species, both animal and plant, are seen as relatives.

It is also important to acknowledge that examples of this approach, although expressed in different ways, also exist in the Western tradition. Albert Einstein's view, coming from the scientific paradigm, was eloquently expressed in a letter to his mathematician friend and classmate Marcel Grossman. He wrote that an individual's feeling of being "separate" is in fact "an optical delusion of [his] consciousness" (Einstein 2015).

Similarly, American biologist and ecologist Edward Wilson developed the concept of "biophilia," around which a whole school of thought has emerged, positing that humans have an innate desire to seek connections with nature and other living beings and that this desire is actually part of biology (Wilson 1984). Another American biologist/ecologist, Aldo Leopold, developed the idea of a "Land Ethic." In his classic book, *A Sand County Almanac*, he explains "the Golden Rule":

> There is as yet no ethic dealing with man's relation to land and to the animals and plants which grow upon it. The land-relation is still strictly economic, entailing privileges but not obligations ... The extension of ethics to this third element in human environment is, if I read the

evidence correctly, an evolutionary possibility and an ecological necessity ... I regard the present conservation movement as the embryo of such an affirmation. (Leopold 1949)

Leopold believed that environmental values were something that grow from experience and a sense of place in nature and with other animals. When exploring various conceptualizations of the relationship to nature and other animals, we can see that some basic premises about interconnectedness are shared across cultures. All of these approaches start with the view that there is an intrinsic value in all life, that we are by nature "connected," while simultaneously rejecting the commodification of humans, nonhuman animals, and nature that we see running rampant in the individualistic, profit-driven consumer world.

> **Vandana Shiva, Author, Activist, and Food Sovereignty Advocate:** That's why I call food a currency of life. And as a currency of life, maintaining that flow is food justice ... The industrial agricultural model asks: how do I extract the most out of a cow, out of the soil, out of a plant? Not even to meet human needs necessarily, but to feed the market and the commodity flow. So flow of life has been substituted by flow of stuff and commodities. And this commodity flow has obstructed the relationships, the interconnectedness, and of course, it has blocked the compassion ...
> So what's the best way to deal with the issue of violence against animals in factory farming? Remove them from the factories. Let them be in family, with you, in small farming families ... For my mother, the animals were her family. That's where compassion and care is taken up in a very natural way. You can't have an extractive economy whose only measure is how do you extract the most and then put a layer of compassion on that. The compassion is in the relationship. (Interview with Vandana Shiva 2023)

Responding to Structural Violence through "Big C" Compassion

With the issues of climate change and relations with other animals and the land in mind, we draw on the spirit of these various traditions. In this vein, all species and all oppressions are in some way connected. We start from a viewpoint of interdependence that understands human beings as both existentially and biologically part of an intricate network of relationships with other human and nonhuman animals and with the earth itself. We argue that harm and violence to one part of the network inevitably affects all other parts, but at the same time, radical, or "big C," compassion towards other human beings, nonhuman animals, and nature can both heal and transform this reality.

What does big C compassion look like in the day to day? As Terry has argued elsewhere, it is a compassion beyond the individual, rather it is expressed through societal structures and processes and in collective goods and projects (Gibbs 2017). It is a structural compassion. It can be difficult to face up to where we cause harm, both as individuals and through participation in particular institutions and processes that often feel out of our control. We will address the related issues of denial, grief, and apathy in the next chapter.

Our interconnectedness is both bad and good news. It's bad news because the so-called "externalities" connected to the food systems that we identify in this book, sometimes referred to as "spillover" – human rights violations; environmental pollution and environmental racism; climate change disruptions; cruelty towards nonhuman animals; destruction of the plant and animal biodiversity upon which our future health depends, etc. – co-exist with so-called democratic societies dominated by political and economic practices that reinforce individual self-interest, overconsumption, greed, and exploitation (Gibbs & Harris 2020).

But we believe that the reality of the interconnection with all other beings and to the earth (reinforced by science) is ultimately good news because it makes clear that what we do as individuals in daily life can affect positive change on the spot; what we do in the everyday matters. Our actions can take place on many fronts including reorienting our perspective and understanding our history in ways that are appropriate, helpful, and healthful to the future of humans, nonhuman animals, and our broader ecosystems.

As the legal entity known as "Canada" struggles with its legacy of genocide of the Indigenous Peoples, we are forced to recognize two truths. Firstly, that Indigenous Peoples lived on these lands for thousands of years in a respectful and sustainable relationship with the land and other animals. And secondly, the wealth and supposed "development" that the privileged and disproportionately European and Anglo-American white settler populations have enjoyed under capitalism on these lands have been rooted in a history of violence, slavery, and genocide that continue in various forms into the present day (Leech 2012; Nibert 2013; 2017).

Some historians have explored pre-industrial Europe, and feudalism in particular, highlighting the enclosure movement as key in the move towards the commodification both of land and labor (Foster & Burkett 2017; Tucker 2020). Foster and Burkett, drawing on Marxist accounts, show how this period marked a fundamental shift in the relations of laborers/farmers with the land and with nature in general. They note, "The bondage seen, both in the feudal order and the ancient world, that linked the labourer with the soil as much as the river with the land, had to be torn asunder to create the expropriated peasant ready to be absorbed for factory production" (Foster & Burkett 2017: 108).

The numerous Enclosure Acts passed by the British Parliament, particularly those between 1750 and 1815, essentially privatized public lands that were previously accessible to all for gathering, hunting, fishing, and grazing animals. The privatization of these lands, known as the "commons," ensured that people no longer had access to the necessities essential for their survival. With no other means of subsistence, they were forced to relocate to cities and become wage-laborers in the newly emerging factories. We argue that a re-capturing of the "commons" is essential if we are to achieve sustainable and humane food systems rooted in democratic principles. But what could a non-feudal commons look like in the twenty-first century? And to what degree does this idea fit with Indigenous knowledge systems and liberation?

The colonization of much of the world by European nations ensured a similar form of enclosure in the colonized territories that displaced Indigenous Peoples from their traditional lands, which were subsequently occupied by settlers. Elder Marshall has suggested that, with humility guiding us and with a history of colonialism still playing out, we must face who we have been, who we are now, and who we want to be (looking seven generations ahead, ideally!) to address these historical harms (Interviews with Elder Albert Marshall, Spring 2022). Being compassionate, creative, and resilient in relations with other humans and nonhuman animals will be essential, he notes, to survival.

Elder Marshall is focused on the importance of decolonizing education to liberate and empower youth, Indigenous and non-indigenous alike. In a talk given at a gathering of Indigenous language educators in 2017, Marshall reflected on the implications of colonialism for education specifically:

> When we lost our connection to our lands, we lost our entire culture. Without our connection to our lands, we had to surrender our children to an education system shaped and created by foreign hands. This education has no connection to the land, and so, in a terrible and quiet way, education has become the new form of genocide. Does this mean that there is nothing of value to be learned from the many generations of accomplishment in other cultures? No. Such a belief would be foreign to our culture. Rather we believe that what is required is to right the balance. The education we taught, the method within nature, the community delivery, the passing on of knowledge by Elders who have had the experience of observing through long periods of time are all principles which were lost and need to be regained. Most important, our connection to the Earth and the fellow species that inhabit this planet must be restored within education. Without this fundamental connection within education, we lose our rights, our culture, our language and ourselves in any manner that has meaning. (Marshall 2017)

Elder Marshall has been working closely with the authors on various community initiatives including the Knowledge, Peace, and Friendship Garden project at Cape Breton University, which focuses on community building and cross-cultural learning through outdoor, land-based education. We share the view that using diverse forms of education (beyond the classroom lecture format), including taking university students and other learners outdoors, is a key way in which we can model the values we are promoting in the world (Kimmerer 2013). As Indigenous scholar Marie Battiste eloquently states in *Decolonizing Education: Nourishing the learning spirit*:

> Education is the belief in possibilities. It is a belief about knowledge systems. It is a belief in the capacities of ordinary humans. We as educators must refuse to believe that anything in human nature and in various situations condemns humans to poverty, dependency, weakness, and ignorance. We must reject the idea that youth are confined to situations of fate, such as being born into a particular class, gender, or race. We must believe that teachers and students can confront and defeat the forces that prevent students from living more fully and freely. Every school is either a site of reproduction or a site of change. In other

words, education can be liberating, or it can domesticate and maintain domination. It can sustain colonization in neocolonial ways or it can decolonize. (Battiste 2013)

In speaking of the communal and experiential aspects of Mi'kmaw education specifically, Battiste also notes: "Cyclical and patterned knowledge from living with nature reinforced in ceremony, tradition, and teachings are important reference points for the knowledge holders" (Battiste 2013).

A similar approach to this type of teaching and learning is recognized in other cultures such as the "popular education" model, also known as "critical consciousness," developed by Brazilian educator/activist Paulo Freire. This approach – using educational situations to empower learners to make positive change in the world – can be applied to how we work in community, recognizing the interconnections between various forms of oppression including poverty and food insecurity.

For this reason, many advocates for land and food justice, such as critical race feminist and food studies scholar Breeze Harper, call for intersectional approaches and language that cut across issue lines in ways that are helpful to understanding different levels and cross-cutting layers of marginality and exclusion. We discuss the implications of her approach to Karen Washington's idea of "food apartheid" more fully in chapter 4 (Interview with Breeze Harper 2022).

Interdependence as a Framework for Advocacy and Social Protest

At one point during the writing of this book, the authors were busy with the local Climate Change Task Force (CCTF), organizing an afternoon of activities that were scheduled coincidentally and ominously – to the day – with the arrival of Hurricane Fiona and with our island literally being in the eye of the storm. The events included a rally designed to call on our government to truly act as though we are in a climate emergency, recognizing the implications for land, humans and other animals.

The signs that we had made for the rally included messages such as: "Every Child Matters: Think Seven Generations Ahead!"; "There is no Planet B"; "Save the Phytoplankton!"; "Decolonize Your Mind!" on a backdrop of a Pride flag; "No More Coal, No More Oil, Keep Carbon in the Soil!"; "Homeland Security: Fighting Terrorism Against Land, People and Other Animals Since 1492" with the backdrop image of Indigenous warriors; "Bee the Change: Save our Pollinators!", etc. What has become increasingly obvious to many is that all struggles for justice for human and nonhuman animals and for environmental harmony are inextricably linked.

Youth perhaps understand these links better than the rest of us. From the science side, Biology student Owen Gibbs Leech (Terry's son) spoke at the rally about the importance of bees. His concerns are echoed by the FAO whose bee experts, as part of a landmark study, noted that over a third of global food production depends on these tiny creatures (Intergovernmental Science-Policy Platform on Biodiversity and Ecosystem Services [IPBES] 2019). Many youths are focusing their attention on protest, education, and changing the legal system. This is highlighted around the world in the Fridays for Future movement started by Swedish environmental activist Greta Thunberg and in initiatives such as the lawsuit currently under way in the province of Ontario, Canada.

A cross-cultural group of seven youth from the environmental law non-profit Ecojustice have brought a landmark lawsuit to the Ontario Superior Court alleging that the provincial government is failing in its climate plan to protect them and future generations. This case marks the first time a climate lawsuit directed towards government policy has had a full hearing in court. Their arguments point to violations of their rights under Sections 7 and 15 of the Charter of Rights and Freedoms, specifically the rights to life, liberty and security and the right to equality under the law without discrimination (Jones 2022). The goal is to get the court to force the province to create a new plan based on science and commit to the goals of the Paris Agreement to achieve targets well below 2 degrees Celsius.

Owen Gibbs Leech, Student of Biology at Cape Breton University: Well, I've learned that the tiniest organisms usually have way more impact than the larger creatures. You could get rid of a large creature, you could get rid of humans, or you could get rid of elephants, and that would have large effects on the ecosystem. But if you were to get rid of phytoplankton or certain small species like bees, it would have disastrous effects on the environment because they come in large numbers and they do things that are very important, like photosynthesizing and getting rid of carbon dioxide and pollinating our food plants, and usually they're much more plentiful. And even though they're tiny, oftentimes collectively they weigh more than all elephants or other creatures combined just because of their quantity. And that's the key because there are so many of them, they can make massive changes and they can allow other creatures to live.

Microbes could live without us, but we could not live without microbes ... Many people do understand that it's important to protect smaller organisms, but I think with the regular population, their idea of giving to charities to help the environment is mostly to protect cuddly things like panda bears, which is important, but saving pandas is far less useful to the environment than phytoplankton or many of the smaller, less cute creatures that are doing way more for the environment, and usually they kind of get forgotten, and are only really talked about in scientific circles and not among the regular populace.

So, I think there's a lack of education and more people need to understand the significance of microbes and creatures like bees. Bees, actually, are an example of something that has caught on. More people do want to protect bees but ... probably because they look cute and fluffy! Definitely, pandas and elephants and other larger animals need to be protected, but we also need to protect the small organisms and creatures too and that relates to questions of broader eco-system protection. (Interview with Owen Gibbs Leech 2022)

Meanwhile, locally, the CCTF is working with Mi'kmaw Elders and youth on drafting an ecological manifesto that incorporates the "precautionary principle." This approach involves not implementing development projects or investments without first assessing possible harm and degrees of harm to land, humans, and other animals with the burden of proof being on developers and investors not on the public (Ongoing dialogue with Elder Albert Marshall and CCTF Coordinating Committee).

In his fall 2022 address to the UN General Assembly, the Secretary General of the United Nations Antonio Guterres warned that even though, at minimum, the world needs emissions to be cut by close to one half before 2030, our fossil fuel "addiction" puts us on track for a 14 percent increase. He spoke particularly to leaders from G20 nations calling for phasing out coal, investing in renewables, and putting an end to this unhealthy addiction while stating, "The fossil fuel industry is killing us."

Meanwhile back at home, Elder Marshall and the authors confronted the reopening of the Donkin coal mine after the last mines and the steel plant on this island closed in 2001, having left this now post-industrial community in a state of serious pollution and with high rates of cancer, disproportionate addictions and trauma, and one of the highest child poverty rates in the country (Frank et al. 2021).

Situating Our Relationship to the Land and Other Species within the Law

It is important to highlight perspectives and experience from various places around the world where nature and/or other species are given, or thought to deserve, "rights" and that argue these rights could or should go beyond "voluntary" practices and be entrenched in both international and national laws so that they are justiciable. At the global level, these ideas are far from new; Principle 15 from the *Rio Declaration on Environment and Development* (United Nations 1992) is instructive for considering what a "precautionary" approach means in practical terms: "Where there are threats of serious or irreversible damage, lack of full scientific certainty shall not be used as a reason for postponing cost-effective measures to prevent environmental degradation."

In their Wingspread Statement on the Precautionary Principle, environmental experts from Canada, the United States, and Europe note: "In this context the proponent of an activity, rather than the public bears the burden of proof" (Wingspread Conference 1998). Others have argued that the actions and policies that do not take this approach often lead to outcomes that can and should, in certain cases, be defined as crimes against nature and other species.

In this vein, efforts have been made in recent years to include the crime of "ecocide" in the Rome Statute of the International Criminal Court. In 2020, the Stop Ecocide Foundation convened an independent panel of 12 lawyers from around the world with expertise in criminal, environmental, and climate law and charged them with preparing a practical definition of the crime of "ecocide." The panel was supported by outside experts and involved public consultations with groups ranging from legal, economic, political, youth, faith, and Indigenous knowledge systems from around the world.

The definition they created reads: "For the purpose of this Statute, 'ecocide' means unlawful or wanton acts committed with knowledge that there is a substantial likelihood of severe and either widespread or long-term damage to the environment being caused by those acts." The most significant terminology for our immediate purposes here is further explained in the definition:

> "Widespread" means damage which extends beyond a limited geographic area, crosses state boundaries, or is suffered by an entire ecosystem or species or a large number of human beings ... "Long-term" means damage which is irreversible or which cannot be redressed through natural recovery within a reasonable period of time. (Stop Ecocide Foundation 2021)

The protection of other species and nature has already been included in the legal systems and constitutions of countries such as Ecuador and Bolivia in response to grassroots movements of Indigenous Peoples and those internationally who work in solidarity with them. Attention was brought to the need for international collaboration and international law specifically after the *Cochabamba People's Conference on Climate Change and the Rights of Mother Earth* held in April 2010 where a Universal Declaration on the Rights of Mother Earth was presented (World People's Conference on Climate Change 2010). This document continues to inform debates around how we may view nature in legal terms.

In Ecuador, a country that has been a leader in this area, then-president Rafael Correa, through a Constitutional Assembly and with the assistance of the "Citizens' Revolution," oversaw the rewriting of the constitution. The goal was to recognize "the inalienable rights of ecosystems" and to entrench this in the constitution of the country. Articles 10 and 71–74 ensure that citizens can petition the government on behalf of nature and that government is required to address and remedy the rights violations. Article 71 reads:

> Nature, or Pacha Mama, where life is reproduced and occurs, has the right to integral respect for its existence and for the maintenance and regeneration of its life cycles, structure, functions and evolutionary processes. All persons, communities, peoples and nations can call

upon public authorities to enforce the rights of nature. To enforce and interpret these rights, the principles set forth in the Constitution shall be observed, as appropriate. (Constitution of Ecuador, Political Database of the Americas 2008)

Our key question is: how can we promote and work towards democratic and compassionate systems of decision-making and food production – two areas that we are sure cannot or, rather, should not be separated – globally that are place-based and culturally appropriate? As we made a deep dive into our research and interviews, we wanted to make visible the barriers preventing us from getting to this place and, at the same time, we wanted to celebrate the incredible things already happening on land and in harmony with other species around the world: in gardens, schools, communities, clean energy development and innovative technologies. It is clear to us that humans have the tools, and many have the will, and/or could be inspired, to build a more compassionate and democratic world.

Our goals in writing this book include the importance of acknowledging how we got here, supporting efforts to stop the worst harms immediately, and working towards better ways of living together on the land and with other cultures and species. Vandana Shiva talks about "decolonizing our minds" in this larger sense, and in her discussion of the "democracy of species," Robin Wall Kimmerer notes that, "In the Indigenous view, humans are viewed as somewhat lesser beings in the democracy of species. We are referred to as the younger brothers of Creation, so like younger brothers we must learn from our elders" (Kimmerer 2013: 346).

The Logic of Capital

While all our problems of disconnection with one another, other animals, and the earth are not reducible to capitalism and its concomitant systems of industrial food production, we situate our analysis within the structural violence of capitalist history and its contemporary manifestations. This is important because this is the dominant system in our world and the one in which all countries and cultures participate.

We are not oblivious to the "progressive" outcomes of capitalism for some, or to the advances in research and technology that have accompanied capitalist expansion, albeit often driven by military and profit motives. We are focused here on what sorts of systems could continue to bring us benefits in terms of social and collective good without depending upon methods that are structurally violent towards so many species.

With this in mind, we will refer in this book to the "logic of capital," a concept rooted in the Marxist critique of capitalism and further developed by many contemporary authors and scholars such as Garry Leech. In his book *Capitalism: A structural genocide* (2012), Leech explains that the logic (and indeed the fiduciary legal responsibility to shareholders) to increase rates of corporate profit leads to outcomes that are inevitably oppressive and often violent at a structural level.

Leech argues that if decision-makers see that certain outcomes are inevitable and they witness particular outcomes on a regular basis – such as inequality, poverty, malnutrition, and death – and yet continue to engage in their economic activity, then we can say there is intentionality and responsibility. He notes: "if a social system creates and maintains inequality in both power and wealth that benefits certain social groups while preventing others from meeting their fundamental needs, even if unintentionally, then structural violence exists. And if such inequality is inherent in a social system, then so is structural violence" (Leech 2012: 12).

Drawing on extensive data from the Global Health Observatory Data Repository of the World Health Organization, his research reveals that millions of humans die every year unnecessarily as a result of processes and institutions that we can directly connect to capitalism. As he points out:

> More than 10 million people globally die each year from hunger and from preventable and treatable diseases such as malaria, diarrhoea, tuberculosis and AIDS, with sub-Saharan Africa the region most seriously impacted. This structural genocide is a direct consequence of acts that adhere to the logic of capital, which ensures food security for the global North and significant profits for agribusinesses and pharmaceutical companies, but fails to see value in human beings whose labour power is not required and who are too poor to be consumers. (Leech 2012: 6–7)

Irrespective of whether one follows Leech to his conclusions about the capitalist system as a whole being genocidal, his analysis of some of the most destructive and seemingly inherent features of the model is in line with that of other well-respected economists, scholars, practitioners, and religious figures such as Pope Francis, Naomi Klein, George Monbiot, Thomas Piketty, and Vandana Shiva. Economist Thomas Piketty, for example, argues that wealth concentration, and therefore inequality – an expression of structural violence – is inherent to free-market capitalism.

One of his central arguments is that the free-market system has a natural tendency towards increasing the concentration of wealth. Piketty's solutions include a global tax on capital and punitive taxes on those earning over $500,000 per year. *The Economist* magazine predictably had problems with Piketty's focus on making the rich pay such taxes. It argued, "Thomas Piketty's blockbuster book is a great piece of scholarship, but a poor guide to policy" (The Economist 2014). Marxists, of course, would point out that taxation of the rich is not in reality about redistribution but is in fact a rather inadequate way of returning to the producers the true value of the products of their labor.

In 2022, the World Bank projected that COVID-19 and worsening inequality would add 198 million people to the ranks of the "extreme poor" during that year, "reversing two decades of progress." Working with World Bank figures, Oxfam was estimating that rising global food prices alone would push 65 million more people into extreme poverty. A 2022 Oxfam International press release gives a statistical picture of what even the relatively moderate changes proposed by Piketty could do:

> Despite COVID-19 costs piling up and billionaire wealth rising more since COVID-19 than in the previous 14 years combined, governments – with few exceptions – have failed to increase taxes on the richest. An annual wealth tax on millionaires starting at just 2 percent, and 5 percent on billionaires, could generate $2.52 trillion a year – enough to lift 2.3 billion people out of poverty, make enough vaccines for the world, and deliver universal healthcare and social protection for everyone living in low- and lower middle-income countries. (Oxfam International 2022)

The fact that even Piketty's modest suggestions to build more accountability into the system are met with such negative responses like that of the *The Economist* demonstrates what we are up against in trying to democratize our economic system and establish meaningful concepts of citizenship and of care and respect for other species and nature.

Corporate Dependency

Perhaps the system we most urgently need to tear down as we work towards radical democracy is the one that allows corporate domination of our democratic systems, the one mechanism we are supposed to have to ensure meaningful control over the decisions that affect our lives. We have discussed above how the logic of capital, in its instrumentality and commodification of everything, is structurally violent. It also permeates every aspect of Western culture down to our daily behaviors.

While neoliberal fundamentalists since Ronald Reagan and Margaret Thatcher have long bemoaned the dependency and passivity that big government creates, they themselves have contributed to the ideological and practical manufacturing of a dependency we could not have imagined under the welfare state. This dependency is not on governments, but on corporations and what they provide.

We have become so dependent on corporations that it is difficult to see a way out. How do we not buy gasoline for our cars? How do we do without a cell phone in today's work and social reality? How do we access affordable, environmentally friendly energy for our households if our corporate providers do not see this as a priority? Unless we are very privileged, how do we buy food and clothing that does not come from unsustainable human rights-violating, and biodiversity-destroying production methods when that is what is cheapest and most readily available? The truth is that most people aren't able to do these things under the current system even if they wanted to.

The wheels of this dependency are greased by advertising and social media, which make everything seemingly accessible at our fingertips (at least for those with a certain degree of privilege). The result is a society of consumer-citizens who primarily express themselves and many of their values through the act of consuming. The consumption mania that has evolved under neoliberalism has promoted ever-increasing levels of individualistic behavior and, conversely, a diminishing of the collective consciousness.

This brings us to another teaching emphasized by Elder Albert Marshall: with rights and privileges come responsibilities. Marshall notes how, in our individualized societies, many have forgotten their responsibilities to one another, to the land, and to other species. We would argue that this is reflected in our broader forgetting as societies of the collective good, both as an aspiration for modern society and as a way to be in the world.

Our corporate dependency, fostered and continuously nourished under high-tech, deregulated, neoliberal capitalism, ensures that we stay forgetful and, therefore, that many of us remain silent about the systems of structural violence – racism, sexism, homo- and transphobia, speciesism, ableism, classism, environmental destruction – that are pervasive in our world. And, as Marshall reminds us, our deafening silence will ensure the continuation of this violence.

The notion of responsibility is also built into the Western rights framework and is implicit in the various UN treaties and conventions that seek to protect political, civil, social, economic, and cultural rights. But as scholars such as Sophie Riley have pointed out, corporations with their industrialized and internationalized practices in the increasingly deregulated neoliberal era, are incentivized, and in fact encouraged, to engage in practices that violate the principle of responsibility for the sake of profit or, as it is known in corporate doublespeak, "prosperity" (Riley 2022).

This is not to say that there aren't corporate CEOs genuinely interested in corporate social responsibility. As former leader of the Canadian Green Party Elizabeth May points out, "There are CEOs and corporate boards attempting to do good in the world, there clearly are. But the notion that there's inherently anything socially responsible about a corporation is a counterintuitive proposition ... by law the corporation exists for one reason: to ensure benefits to shareholders" (Elizabeth May, quoted in the documentary *The New Corporation* 2020).

Any alternative vision of the food system, rooted in the notion of responsibility, will ultimately be dependent upon working towards increased transparency and accountability, and it is our perspective that ultimately these can only emerge with radical democracy.

Towards a Vision of Radical Democracy based in "Big C" Compassion

With this in mind, we advocate here not for depending upon or tweaking existing democratic structures but rather for a structural transformation rooted in transparency and the principles of true participatory democracy. We argue this from the very simple premise that the ultimate meaning of democracy is to give citizens meaningful control over the decisions that affect their lives. As Terry has argued elsewhere (2017), liberal democracy has historically been associated with capitalism; therefore, it has been rooted in the separation of the political and economic spheres of life.

This has meant that democracy exists to varying degrees in the political sphere (depending on what country you are living in) but it is generally very limited in the economic sphere where it tends to be reduced to state regulation (e.g., setting standards around things like food labeling, workers' rights, or environmental regulations for companies) and welfare-based policies. Meanwhile, the major economic decisions that directly impact people's daily lives are made in corporate boardrooms and by unelected technocrats at international institutions such as the International Monetary Fund (IMF), the World Bank, and the World Trade Organization (WTO). Significantly, democracy under capitalism has never extended in any broad-based or meaningful way into the work environments that most of us inhabit for most of our adult lives or to the broader economic system (Held 1995; Gibbs 2017).

Neoliberal policies and the forms of economic globalization promoted by Western powers since the late 1970s have led to a hollowing out of democracy in "advanced" capitalist countries and to the continuing reality of neocolonial relations with the Global South (Interview with Aviva Chomsky 2022). Viewing the terrain of democracy globally makes this problem all the more visible. It does not take an economist to see that some of the core values we associate, however problematically and ideologically, with countries that have long liberal democratic traditions – the capacity to influence the decisions that impact our lives and the right to be treated as equal citizens (i.e., our wealth should have nothing to do with our capacity to exercise our rights) – are not only compromised under capitalism, their violation is part and parcel of the system.

The degree to which there is no real expectation of genuine democratic accountability in the global economy, and the level of political bias in democracy rankings towards the perspectives of the "advanced" capitalist countries and corporate actors of the Global North – passed off as objective analysis – is demonstrated quite visibly in the "Democracy Index" of the Economist Intelligence Unit (2022). While it is not entirely flawed in its approach, this index prioritizes democracy in the political sphere far more than in the economic sphere and there is certainly no suggestion that influential economic decision-makers, such as corporate CEOs specifically, or corporations in general, should be made publicly accountable somehow to the general population in some sort of democratic process. Nor do they acknowledge that foreign supporters of dictators or authoritarians (unless it is Russia or China!) should be much lower on the democracy ranking.

We see those political allies of the Western powers with massive and well-documented human rights violations such as Colombia (now with state complicity in grave human rights violations pre-2016 finally acknowledged) and Israel (permanently engaged in an occupation akin to apartheid and an expansion of settlements in direct violation of international law) are deemed to be comparable to the United States as democracies. Interestingly, as Leech points out, these two countries have for many years been at the top of the list of the largest recipients of US military aid which results in massive profits for US military weapons producers and contractors (Leech 2011; 2012).

At the same time, Cuba, one of the few countries in the world to have achieved the "most sustainable model of development" in both 2006 and 2016, according to the World Wildlife Federation – which uses the UN Human Development Index and the "ecological footprint" to determine its calculations – is deemed to be an "authoritarian regime." And yet, there are many important aspects of democracy in Cuba that we could equate with economic and social democracy that are rarely acknowledged and that do not exist in many, if not most, other countries of the world including the United States and Canada.

In fact, the bias of Western governments and the mainstream media towards Cuba actively ignores this reality and focuses on demonizing the country because its socialist model challenges neoliberal orthodoxy. It behooves us to have some humility and respect for the remarkable achievements Cuba has made in prioritizing democratic access to healthy affordable food, housing, medical care, and university education. As analysts such as Avi Chomsky have argued, Cuba has achieved rates of literacy, infant and child mortality rates, education, and access to health care far beyond other Global South countries and even better than many wealthy nations (Interview with Avi Chomsky 2022). In fact, life expectancy for Cubans is three years more than for Americans; 79 years compared to 76 years (Minto 2022).

Interestingly, when the Economic Intelligence Unit (EIU) accurately shows capitalist countries in the Global South such as Haiti to be authoritarian, those nations' gross inequalities and extreme poverty are never associated with their "capitalist" model of development, whereas Cuba's problems are always linked to its "socialist" policies.

In addition, the EIU talks about "national sovereignty" as the "bedrock of democracy" and yet does not measure the negative impact of neoliberal deregulation on national sovereignty. Neoliberal policies have resulted in a decreasing ability of most countries in the Global South to properly feed their populations because of international rules of trade and structural adjustments and loan conditions imposed by international institutions (Gibbs 2017). Shiva and other women scholars, producers, scientists, and activists from around the world have argued convincingly about the relationship of food sovereignty to genuine or radical democracy, to women's empowerment, and to efforts to combat biodiversity loss and climate change (Shiva et al. 2016a).

As we argue throughout this book, if democracy does not include *the most important/influential decision-makers* in our societies, what does it actually mean? Can it really serve the broad interests of the citizenry? We already know that our current "democracies" in most of the Global North and throughout the world have not been able to effectively uphold or respect the rights of nature or the rights of other species. But, sadly, this is not just about the neoliberal phase of capitalism, it is just more visible and pronounced since the breakdown of the regulatory and protective policies of the post-World War Two Keynesian era (Leech 2012; Piketty 2013; Klein 2014).

The sheer misery and suffering of many (including nonhuman animals), and the poverty and inequality faced by a majority of citizens in the Global South – and increasing numbers in the Global North – under capitalism is a wake-up call for those of us who benefit disproportionately from this system. An analysis of North–South relations suggests that there is an ongoing structural and institutional bias in the global system favoring the cultures and economies of the Global North, whether we examine decisions of the United Nations Security Council or the policies of our major global economic institutions.

The viewpoints, cultural perspectives, and concrete alternatives promoted by many in the Global North and South are often ignored in mainstream political and economic discourse and in the media and/or are actively fought against. A clear example is the responses to progressive alternatives that have emerged in communities and governmental policy in Venezuela, Bolivia, Ecuador, and Nicaragua in recent years (Interview with Avi Chomsky 2022).

As we argue above, when we extend the "welfare of others" to include other species and nature and entrench this in policy and law where possible, our picture of democracy shifts radically, and all kinds of new avenues begin to appear possible. But in order to understand the "hollowing out" of our democracies in the Global North, it is imperative to understand how the framework of international trade and state regulation has shifted in the neoliberal era. It is also crucial, in the context of social justice, to explore both the environmental impacts of this process and the distinctly "neocolonial" ways in which neoliberalism has been implemented in the Global South (Klein 2008; Leech 2012; Gibbs 2017).

Neoliberalism and the Myth of "Development"

While the "economic opening" process that we associate with "neoliberalism" has been happening all over the world over the last four decades, in the Global South it is largely associated with the International Monetary Fund (IMF) and its programs of economic reform, also known as Structural Adjustment Programs (SAPs). This era has been characterized by the transnationalization of economic activity as states transitioned from national to transnational ways of producing and consuming.

This process, according to its advocates, promises prosperity to nations around the world, and to bring the "third world" up to "first world" standards. However, despite decades of neoliberal globalization, so-called "development" still eludes most of humanity. Since World War Two, the construction of the United Nations framework, and the "decolonization" process, arguably not a single "third world" country besides South Korea has become "first world." So South Korea could be considered the exception that proves the rule. Furthermore, South Korea's development was not a result of neoliberal policies, but rather extensive state intervention in the economy that more closely resembled Keynesian regulated capitalism.

The focus of IMF programs is "macroeconomic stabilization," efficiency, and growth. These involve a blend of policies that always include some mix of fiscal austerity to balance budgets, the privatization of state-run enterprises, and a reduction in barriers to foreign investment, in other words: deregulation. And therein lies the rub. While generally presented as part of a neutral and technocratic set of policies in wise fiscal management, the process of deregulation, like all policies promoted by the IMF and the WTO, has very real ideological implications.

For instance, many of the regulatory frameworks established by national governments following World War Two have been a way to protect national populations from harm or suffering caused by economic activity: harm to humans/workers; harm to nonhuman animals; harm to consumers; harm to the environment; and, at a more systematic level, harm to our democratic rights. We could say that within the flawed capitalist system, regulations are one way that populations have attempted to ensure the protection of the welfare and rights of majorities, of other species, and of the environment.

While making the pretense that economic opening and deregulation is about how we can all share our cultures, ways of life, and ingenuity, and that trading with other nations and getting their investments are inherently good things, the reality is that there are only certain actors who can actually take advantage of the kind of opening that capitalist development in a deregulated framework entails. This is an area where Left and Right politics often converge. While from very different angles, many right-wing/leaning and left-wing/leaning protests in recent years, we would argue, reflect a sense of a loss of democratic control and accountability (i.e., Brexit, the election of Donald Trump, Canada's truck caravan during COVID-19, and the anti-globalization and Occupy movements are just a few recent examples).

In reality, and indeed often in law, the focus of economic "opening" is in the interests of economic actors, and the "rights" are for the largest and most powerful of these actors. A very disturbing aspect of this process comes into focus with the expanding legal conceptualization of the corporation as a "person" with rights. This was highlighted in the Citizens United decision of the US Supreme Court in 2010 which gave corporations full rights to spend money how they see fit in federal, state, and local election campaigns.

The separation from democratic oversight of any "development" activities with significant impact should be, and we believe is, of concern to all citizens in a "democracy," no matter what their political colors. A stark example of this reality was inadvertently acknowledged in a recent meeting one of the authors had with the human rights representative at the Canadian Embassy in Colombia as part of an interview conducted by a Canadian election observation team.

The young man, well versed and published on topics of corporate social responsibility (CSR), had visited communities where humans and nonhuman animals have been displaced and/or are facing pollution of their land and water, destruction of biodiversity and pollinator habitat, and/or are facing malnutrition and death due to pollution and diversion of water supplies from communities by mining companies (Human Rights Watch 2022). He noted that it is hard for communities to "prove" that it is the fault of the corporations and without that proof the embassy cannot act.

Terry pointed to the recent well-documented case of almost 5,000 Indigenous Wayúu children who died from malnutrition in the La Guajira region due primarily to a lack of water to irrigate their food crops (López Zuleta 2015). According to Human Rights Watch, in addition to government corruption and climate change, "Colombian high courts have found that mining in the region has also degraded the quality of and access to water for some Wayúu communities" (Human Rights Watch 2020). The principle mining operation in the region is the foreign-owned Cerrejón Mine, the world's largest open-pit coal mine, which provides coal to Canada, the United States, and Europe for power generation.

Mining activities have also forcibly displaced Wayúu and Afro-Colombian communities from their homes and lands. And while reports about this human rights crisis have documented the effects of mining activities not only on humans but also on nonhuman animals in the region – including birds and pollinators and fishes in local rivers and the overall biodiversity of local communities – these concerns are not even on the radar of the embassy with regard to "rights" and they are certainly not on the agenda of investors (Chomsky et al. 2007).

As the Canadian Embassy representative, in a very transparent exchange, made clear, the human rights portfolio and the trade portfolio have distinct and often opposing agendas. He explained that the trade mission has the explicit goal of "increasing the prosperity of Canadians," while the human rights portfolio has responsibility to investigate and report on the human rights situation of Colombians.

The disturbing fact is that there is an acknowledgment of harm to Colombian citizens, but this was by necessity a secondary concern in trade relations, and harm, in general, to humans, nonhuman animals, and nature is seen as inevitable. This reality is simply reinforced by the fact that this corporate activity is completely legal.

> **Aviva Chomsky, Professor of Latin American History and Activist:** As you know, I'm also involved with labor issues in Colombia and the repression of labor. When we were planning some events connected with the Cerrejon mine, Garry [Leech] brought up the idea that we should invite the labor union to a conference that we were organizing. It was like, "Wow, that's really a long shot. But I said, sure, go ahead, let's do it." And sure enough, the union president was really interested in the conference.
>
> The idea of the conference was to create a forum for the voices from the communities who were protesting the mine, and to hold the event right there in the mine's backyard. The new president of the union came to the conference and afterwards he approached me and Steve [Striffler] and Garry [Leech] and said, we have to do something about this, like this was just incredible. And the four of us went out for a drink at the end of the day, and there was this kind of euphoria, like WOW, we have to do something, this is so amazing! And it really was later that I started thinking of it in terms of, wow this is a coal mine, this is a climate issue. And then kind of moving from there and thinking about fossil fuels and capitalism, which as you know, is a lot of what I'm thinking about now ... of course, environmental destruction is inherent to capitalism, but fossil fuels are also really inherent to capitalism.
>
> In my work on the history of the region [La Guajira], I have looked at how the Indigenous populations there remained autonomous and the different factors at the end of the twentieth century that started undermining their autonomy, climate change being one of them. So the communities there are affected not only by the direct pollution from the coal mine, which is what they think about most, but also from the drought that has been plaguing the region and ... it's happening in all kinds of ways. There is the destruction of the water table, as the mine has dug down its use of water, as part of it, but changing weather patterns are also the result of climate change. And I just

> feel like those communities affected are kind of emblematic of third world communities who have contributed nothing to the problem and yet are its primary victims. (Interview with Avi Chomsky 2022)

How is it possible that so many of us, particularly in the privileged sectors of the Global North and South, are ambivalent about an economic system that has created so much harm? This harm has occurred historically through colonialism and in contemporary times with its ongoing legacy that has entailed and continues to entail so much cruelty and neglect of both human and nonhuman animals. And, for the purposes of this book, and given that food is a basic necessity of life, it is a system that, despite its advanced technology and innovation, has failed to feed the world or to nurture a sustainable food system overall.

The Rights of Future Generations

As previous historical struggles to end slavery, to promote civil rights, and to dismantle colonialism make clear, changing history involves both everyday gestures of humility, kindness, and compassion by ordinary people while simultaneously seeking change to policy and law and broader social structures of oppression and violence. In these struggles youth are vital to our visions of the future for many reasons, not least of which, as Thunberg so famously reminded us, sometimes we older folks just don't listen! Not only did our children not create the mess we are in, but they are growing up in a time where eco-anxiety is just a part of the fabric of everyday life; many youths cannot imagine a time in which this was not the case.

Additionally, many of today's youth are, arguably, in many locales, not as burdened by attachment to old ideologies and "isms" around race, class, gender and gender identity, sexuality, disability or a sense of entitlement to dominion over other species and nature, that die hard in the older generations – even for those of us who have spent a lifetime reflecting, working, teaching, and researching in these areas!

But what is clear to us as researchers and teachers who have traveled, lived, and worked cross-culturally and around the world, and who are parents of blended and culturally mixed families ourselves, is the almost universal compassion that children display, without thinking, towards other living beings and to nature in general, before they are "educated" and socialized to believe and behave differently.

Owen Leech, the young scholar interviewed for this book, explains that scientists have long explored the "positive" physiological and arguably equally innate human characteristics that have allowed us to survive for millennia on this planet, including the need for belonging, cooperation, compassion, and an ability to think into the future, traits also identified in the animal world (Interviews with Brandon Keim 2022 and Owen Leech 2023). Leech notes that the popularization of "survival of the fittest" reductionism (emphasizing the dominance of fear, self-preservation, and unavoidable harm and violence) is an incomplete understanding of our history and of our interdependence with all living beings and nature. He rather casually jokes that we should explore what some have called "survival of the good enough," acknowledging both the darker and enlightened sides of the human species.

From an early age Leech, probably like many other kids, questioned the words "nature" and "environment," not sure how these things could be separated from "human" and "alive." He shared that he could sense from a young age the subject/object duality of even our most progressive discourses firmly entrenched in Enlightenment paradigms. This curiosity led him to a passion for both biology and history, the rigid disciplinary separation of which continues to both fascinate and puzzle him, since both are about life on this planet.

What is interesting is that this same disconnect from the language of pure science and rationalism is reflected in Indigenous ways of knowing (see the works of Indigenous scholars Robin Wall Kimmerer, Margaret Robinson, Marie Battiste, Taiaiake Alfred, and many others) and in the understanding of what life is for many cultures around the world (the works and activism of Indian physicist Vandana Shiva, and those of founder of the Green Belt movement, Kenyan Wangari Maathai, come to mind; Shiva 2005; 2016b; 2016c; Maathai 2009; 2010). It is not about being "primitive," "underdeveloped," or lacking in "civilization," as the hugely violent practices and discourses of colonialism continue to try and have us believe.

The history of human beings, Leech argues, inadvertently echoing Robin Wall Kimmerer and others, is inseparable from the history of plants and other animals. In the *Democracy of Species*, Kimmerer notes: "It was the bees that showed me how to move between different flowers – to drink the nectar and gather pollen from both. It is this dance of cross-pollination that can produce a new species of knowledge, a new way of being in the world. After all, there aren't two worlds, there is just this one good green earth" (Kimmerer 2013: 47).

Leech and other youth have also reminded us that we must not succumb to "climate defeatism." For Leech in particular, learning to live in peace and friendship with other beings and the land just makes sense given our interdependence with all other species, and that our safeguarding of plant and animal diversity and balance is not a subject-to-object relationship but rather a subject-to-subject relationship (Interview with Owen Leech 2022).

That philosophical discussion, however, is beyond the scope of this book. Our key aim here is more practical and modest. We aim first and foremost to stimulate a sense of agency in our readers, a sense that no matter where they find themselves as students, parents, tradespeople, teachers, journalists, astrophysicists, or hairdressers (for an impressive campaign on hairdressers promoting climate change awareness see the Sydney Australia movement led by Paloma Rose Garcia), they are influencing history right from where they stand today.

And, whether people intentionally work towards specific goals of social, cultural, and other species inclusion and environmental harmony, or they simply live on auto pilot maintaining the status quo, they are part of a social process that contributes to upholding particular social values. There is no position of neutrality; we are re-enforcing certain values either consciously or unconsciously.

As Elder Marshall notes, our silence is a message of complicity with what is. If we do something, we are contributing to history; if we do nothing and passively follow the status quo, we are also contributing to history; there is no way out of it! This process begins with self-awareness, understanding where we may cause harm or contribute to violence at a personal level, sometimes unknowingly, both through our habitual patterns and by participating passively in broader social systems that are structurally violent (Gibbs 2017).

For those with privilege, recognizing the significance of our own piece of the puzzle both in terms of harm and in terms of potential to influence change can be enough motivation to live more intentionally, maybe to engage in advocacy or activism. But to others it can feel overwhelming and lead to anxiety, depression, avoidance, and sometimes denial. For those just scrambling to get by, the idea of living differently "intentionally" is often not a real "choice."

Filipina youth activist Anya Graf reminds us that making a "choice" to live an alternative lifestyle in our consumer decisions, and in our food choices in particular, is often an expression of privilege, which is not realistic for underclasses in the Global North and for most in the Global South (Graf 2021). In other words, expecting individual consumer behavior alone to change the world is at best naïve and at worst a denial of the real problem, which is a system that doesn't work for most of humanity. To make healthier and more humane choices affordable we must confront larger economic and political structures.

Our current systems of education, media, and economy, as well as our largely undemocratic work environments, tend to reinforce hierarchies and power that breed acquiescence and passivity, insidiously erasing our sense of agency, our joy, and the joy of other animals. At its worst we become completely incapacitated, no longer believing that change is possible, and certainly not believing that we as individuals can make a difference.

Where Do We Go from Here?

Basing our arguments on years of living, teaching, researching, and activism on social justice, animal and environmental issues, and on interviews with several leaders and participants in the movements associated with these ideas, we argue for three very simple things. Firstly, using the lens of "structural violence," echoing the views of Indigenous friends and colleagues, we argue for an immediate and humble acknowledgment and transparency around the harm our global systems of production and consumption have caused historically, and are still causing today. We argue that there needs to be a willingness to simply stop some of these activities immediately, despite the ways in which some of us individually or culturally benefit.

Building on the ideas first advanced by peace scholar Johan Galtung in the late 1960s, and further developed by many contemporary authors such as Garry Leech, we explore how systems with their own instrumental rationality, often unintentionally, can and do ensure the continuation of violence both on a daily individual level and through institutional processes connected to economy, education, and media. Leech argues for the "recapturing of the commons from capital" (Leech 2012: 114). This is perhaps a Western way of framing the Indigenous concept of "right relationship" with the land. We will further discuss the idea of the commons in the final chapter.

Secondly, we promote an appreciation and celebration of all the work that is already being done around joy and gratitude for life and living beings all over the world. There is a multitude of expressions of alternative ways of living on this planet that are compassionate and culturally appropriate to place, and spiritual or religious traditions; we want to highlight the significance of this message to humans around the world.

Thirdly, and perhaps the most practical goal of our work, is the promotion of a radical democratization of our systems of production and consumption, giving people meaningful control over the decisions that affect their daily lives. All these ideas imply new ways of organizing our workplaces, education systems, our politics, and our economics (already reflected in communities and experiments around the world) and, we believe, will by necessity create a more peaceful world.

Each of us will find ourselves with varying degrees of ability to have influence in one or more of these systems based on our identities, resources, education, position, and so on. We will profile several individuals and movements that model new ways of being that make a difference in the immediate moment and that point to more compassionate, inclusive, and democratic societies in the future. But first, we explore the barriers to food justice by looking at our industrial food systems in the context of climate change.

2 Capitalist Dreams and Nightmares

Food Systems, the Animal-Industrial Complex, and Climate Disruption

The solution is not just to reduce your individual consumption, it's to challenge the system that relies on our consumption and consent to keep it going.
– Avi Chomsky

Yet while tyrants since the time of Pharaoh and Alexander the Great may have sought to sit atop great civilizations and rule them from above, never before have our society's most powerful players assumed that the primary impact of their own conquests would be to render the world itself unliveable for everyone else.
– Douglas Rushkoff

You don't need a PhD in Economics to know your life sucks under capitalism.
– Kshama Sawant, in *The New Corporation* documentary

Denial as the First Stage of Grief

It has been suggested that grieving is part of the research process and is ultimately a political act. From a feminist perspective, authors such as Kathryn Gillespie and Patricia Lopez argue that often, as academics and practitioners, those of us who are witness to violence and loss in human, other animal, and ecological contexts are expected to have little or no emotional connection to our subjects of study and engagement (Gillespie & Lopez 2020). We also often see an emotional disconnect, or lack of empathy, in elected political representatives. However, there are increasing numbers of politicians frustrated by this state of affairs. Kshama Sawant, an Indian-American politician and economist who has served on the Seattle City Council since 2014, expresses this sentiment in the quote above, taken from a panel discussion with filmmaker Joel Bakan. The work of women such as these asks us as educators and decision-makers to reflect on how we personally relate to humans and nonhuman animals on the lands where we work.

Douglas Rushkoff, professor of media theory and digital economics at Queens/CUNY, and host of the podcast *Team Human*, has personally engaged in discussions with some of the world's most powerful economic players. He talks about the "incongruence" many of us feel when we are acting in our professional roles and trying to be "balanced" in our sharing and delivery of the hard messages. For him it is a matter of principle that we allow how we really feel to be present in our ways of being with one another whether in professional environments or not (Rushkoff 2022). In our incongruence, he suggests that we do a disservice to ourselves and to others.

It is appropriate to express grief at what is happening to our world. Australian Geographer Lesley Head, who specializes in human–environment relations, writes in her recent book *Hope and Grief in the Anthropocene*:

> an underacknowledged process of grieving – with all its complexity, diversity and contradiction – is part of the cultural politics of responding to climate change and associated environmental challenges. I argue that grieving helps explain the denial we face and experience in accepting the scale of the changes required in ways of living ... I reject the cultural assumption that even to canvass these issues is to give in to them, to give up or to assume the worst. Rather I argue that a relentless cultural disposition to focus disproportionately on positive outcomes is itself a kind of denial. (Head 2016: 1–2)

Politics enter the arena when we begin to discuss what to do about our grief. Head argues that being in tune with possible negative scenarios is more honest and ultimately may allow for more creative responses: "If, for example, projected temperature increases for later this century span an envelope between 1.8C and 4C, we have no scientific basis for planning for the lower figure just because we like it better and wish it so" (Head 2016: 168). Understanding that we cannot "plan" for every possible outcome and that some situations may be "ungovernable" could allow for more expansiveness as we navigate the creative possibilities of what it means to be a citizen of the Anthropocene. For Head this involves a future beyond the modern, Enlightenment view of linear progress.

Other researchers such as sociologist and environmental scholar Kari Marie Norgaard have explored the political spectrum of climate denial, highlighting elements of fear and impotence:

> We can see these forms of denial on a continuum not only in terms of causation, but also effect. Emotions of fear and cultures of cognition regarding the end of the modern capitalist system or "life as we know it" have generated climate denial of the far right and a combination of green capitalist technological fixes and climate apathy on the left. Both responses to climate change reinforce existing social structures and solidify power relations at the top. Widespread apocalyptic framing of system collapse on both the left and the right has captured and derailed public engagement either through a deer in the headlights version of public apathy, or through outright rejection of the climate science. (Norgaard 2019: 439)

It is precisely because it is about our survival that the politics around climate change have manifested in denial and polarities. What has become increasingly clear to many of us working in social and environmental justice is that when we fully grasp the implications of anthropogenic climate disruption, we find that food justice, and all the social integration networks this requires, is at the base of our future survival strategies.

In exploring this reality, Norgaard and others draw on the work of professor of politics and global affairs Frank Fischer who points out that climate change denial is much more complicated than proving arguments through "facts." Even if we don't like what the deniers are saying, he notes:

> the socio-cultural logic of practical reason entitles them to their position. The controversy thus does not altogether rest on the climate numbers, despite claims to the contrary, and, as such, is not a matter that can be resolved by better fact checking. As we have seen, it is concerned as much or more about underlying questions concerning modern society, its social values, normative orientations and policy goals. Toward this end, the climate deniers emphasize these implications, with the numbers mainly serving in various ways – intellectually and emotionally – as proxies for these deeper issues. (Fischer 2019: 148)

We believe it is in the appeal to compassion, community resilience, and strengthening human rights that we may reach towards values shared across the political spectrum. We will return to this notion in the final chapter. For our purposes here, we focus on the question of how the appeal to human rights and freedoms, specifically as they relate to capitalism, may be an avenue for change.

The Relationship between Human Rights Protection, Peace, and Food Justice

The United Nations Climate Action argues that protecting biodiversity is a human rights issue. On October 8, 2021, the United Nations adopted a resolution recognizing the human right to a clean, healthy and sustainable environment. This right is now recognized by 155 countries and it points out:

> The Earth's land and the ocean serve as natural carbon sinks, absorbing large amounts of greenhouse gas emissions. Conserving and restoring natural spaces, and the biodiversity they contain, is essential for limiting emissions and adapting to climate impacts ... The main driver of biodiversity loss remains humans' use of land – primarily for food production. Human activity has already altered over 70 percent of all ice-free land. (UN 2022)

In a similar vein, legal experts Carter Dillard and Jessica Blome speak to the idea of freedom entrenched in the US Constitution and the right to live your life without the interference of others (Dillard & Blome 2022). Losing your rights to a clean and healthy environment, because of the actions of other people or entities, clearly needs to be considered in this framework.

It is clear that inclusive and equitable access to fertile land, clean air and water, and nutritious food are fundamental parts of a successful approach to food justice, but they are also rights issues in very concrete and practical day-to-day ways. Science tells us that the possibility of this healthy environment is derived from and depends upon unpolluted soil, clean waterways, and biodiverse growing environments where multiple species coexist in balanced systems governed by laws of nature. With this in mind, there is increasing research identifying the importance of "nature-based solutions" (NBS) to climate change.

As part of a research team with the Nature-based Solutions Initiative of the Department of Zoology, the Environmental Change Institute and the School of Geography and Environment at the

University of Oxford, Nathalie Seddon et al. note: "NBS are gaining traction in international policy and business discourse. They offer huge potential to address both causes and consequences of climate change while supporting biodiversity and thereby securing the flow of ecosystem services in which human well-being depends" (Seddon et al. 2020: 9).

This message ties in neatly with the warnings that Indigenous Peoples have been giving for generations, and that activists and scientists from all over the world are now affirming: without a strategy that begins with an acknowledgment and respect for our relationship to the land and our interdependence with nature and other species (reflected in NBS to climate change), we may not survive.

Before the arrival of the complex production systems of industrial capitalism, living within the laws of nature ensured our future survival, and recent research confirms the significance of this reality for us today. A team of evolutionary biologists and ecologists from universities in France and the United States argue, "Although the need to address the environmental crisis, central to conservation science, generated greener versions of the growth paradigm, we need fundamental shifts in values that ensure transition from a growth-centered society to one acknowledging biophysical limits and centered on human well-being and biodiversity conservation" (Martin et al. 2016: 6105).

As we have already argued, food justice is related to our access to and collective responsibility towards the land, but it also speaks to our capacity for peace in the world. As the United Nations Security Council and other international bodies acknowledge, most wars of the future will be about access to resources such as fertile land and clean water. Using computer modeling, researchers from the European Commission's Joint Research Centre (JRC) revealed in a 2018 report that conflict over water, for example, is likely to emerge in the coming years in several "flashpoints" across the world where "hydro-political" issues exist. They identify the top five river systems at risk of future water conflict: the Nile, Ganges, Indus, Tigris-Euphrates, and Colorado rivers. They warn of a 75–95 percent chance of more water wars in the next 50–100 years (Farinosi et al. 2018).

Arguably, land and resources have always been at the root of most conflicts historically, but what is different today – demonstrated through ongoing responses to worsening climate disasters around the world – is that access to resources, even breathable air, may increasingly depend upon where one lives and one's privileged position within one's own country and within the global system in general.

Privilege manifests as part of climate denial and as a factor in climate resilience. Privilege will be demonstrated and managed through control of everything from transport systems and power grids to education and health-care systems. As Douglas Rushkoff points out in his recent book *Survival of the Richest*, global elites are the first to recognize this in their busy planning of investment and "escape" strategies, running the spectrum from investment in lab-grown meat and survivalist bunker communities to research on the viability of living on Mars (Rushkoff 2022).

Nature in some form will go on without us and, as some analysts suggest, may be relieved to do so! But, as human animals who love this planet, and the other species we share it with, we would like to imagine a future that includes all of us; a future that protects the integrity of the planet for coming generations of humans and as many nonhuman animal species that we are still able to save at this point.

We realized in writing a book about food justice that nonhuman animals, and particularly those in the food industry, may not be a top priority in the minds of most people. Indeed, raising the issue of nonhuman animal lives, their sentience, and their emotional lives, may at best seem "sweet" but not an immanently important or practical concern given the realities of global poverty and inequality, ongoing wars, and increasing climate disruption. At worst there is active denial and even vilification of those attempting to put other species, most notably nonhuman animals used in the production of *food*, on the list of beings that we must think about if we care about our future on this planet and if we want to build more inclusive and compassionate societies.

Deep within the psyche of even the most alienated of us in advanced capitalist societies is the idea that our children will live at least as well but hopefully better than we did, that they will be cared for and be happy and fulfilled. As several recent studies are showing, we are at a point in history where this is now no longer the reality for the large majority, even in the most affluent of societies such as the United States (Cramer et al. 2019). We are arguably entering a period where "hang on to what we have" may become the new political norm.

This reality is expressed in movements across the world from Brexit and the Donald Trump phenomenon to right libertarian expressions of white supremacy and elite community bunker building. But these expressions are only the most blatant tip of a very dangerous iceberg that many of us in the advanced capitalist countries, and we include ourselves in this, are insidiously supporting through our participation in a fossil fuel-addicted, consumption-driven society. Our broad societal denial of our interdependence with other species and the land and, perhaps most symbolically, our passive acceptance of "business as usual" could be the death of us.

It is imperative to acknowledge how grief is being expressed *en masse* around the world. One expression of grief is evident in the farmer suicides that are occurring in many countries including Canada and India (Interview with Joe Parish 2023; Research and Documentaries of Palagummi Sainath). In India, there are close to 30 farmer suicides every day, and hundreds of thousands of farmers have taken their lives during the neoliberal era. If agricultural laborers are added, the numbers are much higher (Shivji 2021).

Many of the farmer suicides in India are a consequence of unsustainable debts resulting from farmers purchasing patented seeds and other inputs from multinational corporations such as Monsanto and Cargill. When these seeds fail to produce yields to the promised levels, farmers are faced with losing their farms due to the money they owe and, consequently, many see suicide as the only solution to their shame and despair. In a cruelly ironic twist, many of these farmers kill themselves by drinking chemical pesticides, one of the agricultural inputs that contributed to their indebtedness (Leech 2012; *Nero's Guests* 2009).

Farmers in the Global South lived for millennia producing food through traditional practices grounded in local communities and serving local markets. This ensured that a basic level of food security was built into the production system. Seeds were essentially part of the "commons," free and available to all. Vandana Shiva points out that this was a problem for seed corporations like Monsanto:

> To create a market for seed the seed has to be transformed materially so that its reproductive ability is blocked. Its legal status must also be changed so that instead of being the common property of farming communities, it becomes the patented private property of the seed corporations. (Shiva 2005: 91)

In this process, corporations benefited from the property rights provisions negotiated during the 1986–94 round of talks of the General Agreement on Tariffs and Trade (GATT, since 1995 the World Trade Organization, WTO). Leech explains that this process of linking intellectual property rights to seeds allowed large agribusinsses to obtain the patents they needed to create genetically modified (GM) and hybrid seeds. The corporations, he notes, do not "invent" these seeds, they take existing seeds and modify them. This has led to a movement of "biopiracy" whereby corporations can make profit from seeds while farmers gain nothing from generations of Indigenous knowledge concerning medicinal plants, hybrid seeds, natural pesticides, and the maintenance of diversity (Leech 2012).

The work of ground-breaking investigative journalist Palagummi Sainath, particularly his award-winning documentary *Nero's Guests*, provides a bleak picture of farming in the neoliberal era and the role of large corporations such as Monsanto in helping to create an environment in which small farmers cannot afford inputs and cannot effectively compete (*Nero's Guests* 2009). His research and interviews provide staggering statistics and painful family stories from around India.

Sainath, former Rural Affairs Editor of *The Hindu*, has received accolades around the world for his work. In New York, he received the Harry Chapin Media Award in 2006 and, for his "passionate commitment as a journalist to restore the rural poor to India's national consciousness," he received the Ramon Magsaysay Award for Journalism, Literature, and the Creative Communication Arts in 2007.

Terry sponsored a visit by Sainath to Cape Breton University in 2007 to speak about a photo exhibition of his work entitled "Women at Work," visually telling the stories of women farmers and their families. Sainath, like Vandana Shiva and other community-engaged advocates interviewed for this book, is an example of what compassionate citizenship can look like, not because we can all perform at this level, but because telling the truth, with a deep compassion for those who are suffering, is a gift to all of us.

Av Singh, Agronomist, Educator, Farmer: In terms of my work, the most challenging thing would typically come down to the fact that the Academy and the federal government don't support the alternatives, and I say alternatives because they're not the dominant practice in North America. They don't support those strategies, despite the fact that they would be economically better and environmentally better. And, of course, the Academy is often pursuing a funding program that is typically more corporate or industrial focused. And the antithesis to that, the successes and the excitement, comes out of farmer-led research, farmer-shared knowledge.

So, being able to watch farmers share amongst each other, building that social capital around knowledge exchange, that's always been incredibly rewarding. My continued passion to work in agriculture comes from farmers who are just constantly learning and making their own paradigm shifts in their whole farming process, whether it's transitioning to organic, or them turning to be more focused on biodynamic farming, or ecological farming, or regenerative farming, or natural farming, all these other things that they just dive into, that's definitely where the rewards are. (Interview with Av Singh 2022)

From Grief and Denial to Confronting Systems

With climate change and disruption as a framing lens, we will now explore the Animal-Industrial Complex (A-IC), situating it as part of the broader capitalist industrial food system to determine how we got into the mess we're in and what the path towards a more just and ecologically responsible food system may look like (Noske 1997; Twine 2012). It is important that we untether ourselves from the sinking ship, not through individualist bunker strategies or protecting our own access to the ship at the expense of others – human and nonhuman – but through more collective systemic changes.

The first step of this process is determining the barriers preventing us from getting there. One of the most significant barriers we face in multiple realms is arguably denial. As journalist and former war reporter Dahr Jamail notes, even those of us very worried about the state of the planet have perhaps been in some denial about where we really are and what this may mean. This is at least partly due to the grief this recognition entails, but also because we may fear that we are ultimately impotent in the face of powerful global forces backed by deeply entrenched ideological systems and military might (Jamail 2020).

What does this denial look like at the level of our relationship with other animals? Even those on the progressive/left end of the political spectrum have often been avoiding, and sometimes actively denying, the significance of other species to both the climate and social justice movements (see Amy Fitzgerald's *Animal Advocacy and Environmentalism* [2019] for an innovative look at the most common disagreements and the important common goals). This is occurring despite the increasing evidence from the scientific community about the disproportionate role of the A-IC in climate disruption and human rights violations (not to mention the billions of nonhuman animal rights violations that are largely legal and accepted).

As researchers relatively new to the scientific studies ourselves, we see that the vital contributions of wild species in carrying out "ecosystem services" and ensuring biodiversity, and even in protecting us against climate disruption, are poorly understood. And while recognizing the "services" performed by other animals that benefit us is not the most ethical reason for treating them well, it is certainly a very practical and intelligent one. As unpopular as the message may be to some, what we want to make clear is that tackling social justice issues – including poverty, inequality, racism, sexism, homo- and transphobia – is inseparable from working towards the liberation of other species from oppressive systems, lifestyles, and attitudes.

Sociologist David Nibert argues that challenging the oppression of other animals, and indeed challenging the destruction of all that is alive, is a fundamental pillar in the overall struggle for a more compassionate and just food system and ultimately a more just world. Nibert has long argued that the use and abuse of other animals by humans has been intimately tied up with the history of colonization and capitalism and the abuse of other – disproportionately but not only – non-white humans, particularly Indigenous Peoples and many in the Global South. For him, and many others, confronting this reality is a starting point both for our critique of current systems and for our construction of a more inclusive and compassionate food system (Interview with David Nibert 2022).

David Nibert, Professor of Sociology and Critical Animal Studies Scholar/Activist: I argue that with the advent of colonization, Europeans never could have successfully invaded and dominated so many people around the world had it not been for their use of other animals as instruments of war and as rations. And even then, it was really the zoonotic diseases brought by the invaders, diseases like smallpox created through the oppression of other animals, that truly devastated many Indigenous Peoples around the world. These terrible epidemics destabilized Indigenous societies and made the European conquest possible as the assailants thundered in on the backs of exploited horses and committed genocide and enslaved people. Importantly, so much of the land that was stolen during colonization was used for the expansion of ranching.

However, in what is now Canada and the United States, plunder and colonization were initially driven due to profits that could be accumulated through the killing of other animals for their skin and hair, or their "fur." The lucrative trafficking in the skin and hair of other animals like beaver and deer bound many Indigenous societies to the ominous rise of the global capitalist system. And it ended up being devastating for Native American communities in several ways, including the development of widespread warfare between Indigenous societies, including the Beaver Wars. And when Indigenous Peoples were no longer useful resources in North America, the colonizers accelerated their expropriation of land for ranching, which was then already pervasive throughout Central and South America.

More and more land was taken for ranching oppressed other animals because this was a primary source of wealth and power, then as now. The fates of such oppressed other animals as horses, cows, sheep, and pigs was deeply entangled with the fates of Indigenous Peoples who faced continual encroachment. Enormous amounts of land were stolen during colonization for ranching, not just in the Americas, but also in Australia, New Zealand, and Africa. And very importantly, to this day much of this stolen land remains in the hands of ranchers and feed crop producers, malevolent use motivated by the existing destructive and unjust global food system. Here in the United States, for example, Indigenous Peoples continue to be devalued and marginalized, while their ancestral land is controlled by powerful ranchers and feed crop producers who long have used their economic and political power to ensure their interests are

> represented in their state houses and in Washington; and their oppressive practices are protected and supported with taxpayer dollars. For example, in the United States ranchers can graze oppressed cows on public lands despite the resulting environmental damage and the squandering and polluting of dwindling supplies of freshwater.
> The same thing continues throughout this hemisphere and in parts of Africa, Australia, and New Zealand. In Brazil, for example, there's massive rainforest destruction caused by ranchers. And enormous areas of the Amazon also have been expropriated by feed crop producers whose soybean meal makes possible the horrors experienced by billions of other animals around the world confined on factory farms. So, the global food system today is a continuation of the violence, oppression, and trauma that began with colonization. This oppressive food system results in the ongoing oppression of Indigenous Peoples in the United States, and throughout the world, through their marginalization, impoverishment, and other forms of institutionalized discrimination. Unfortunately, when most people sit down to eat they have no idea about the consequences and true costs of that meal. This lack of knowledge is part of the powerful ideology that keeps this whole, oppressive system going. And then, I have to say, that there's no way that any of this can be disconnected from capitalism, which has been driving all of this from the start. Capitalism is rapidly driving the world towards environmental collapse. (Interview with David Nibert 2022)

Through systematic historical research, Nibert has traced the history of capitalism through this lens, showing how this system has depended since its beginning on exploitation and cruelty and violence towards humans and nonhuman animals (2013). Nibert goes on to explain how investors saw early on in the colonization process the incredible potential for profit making from ranching, expanding their activities and transport of animals from the plains across the United States to places like Chicago for slaughter, leading to the construction of extensive transport systems of roads and railways built by Chinese and Philippine laborers and Irish immigrants fleeing famine and poverty back home.

This process could not have happened without massive access to public land and what Vandana Shiva describes as "cowboy capitalism" in its robbery of the commons, its exploitation of land, and human and nonhuman animals. This system, Nibert notes, continues to this day with ranchers working hand-in-hand with large banks and Wall Street, the trucking industry, and massive slaughterhouses.

At a day-to-day level this reality is supported and reinforced through various socialization mechanisms including the education system where children from a very young age are "educated" about the need to eat *meat, fishes, eggs* and *dairy* to stay healthy. The growth of the "junk food" market added yet another dimension to the opportunity to make massive profit from the exploitation of land, people, and other animals and is tied in with the concomitant need to keep the public blind to the consequences of this system for their health and the health of the planet they call home (Nibert 2002; 2013).

Numerous documentaries in recent years, from *Food Inc.* (2008) and *Speciesism: The Movie* (2019) to *Cowspiracy* (2014) and *Seaspiracy* (2021), have shown the links between the often invisible, authoritarian practices of capitalist industries and their penetration into the political process through lobbying and campaign donations and the relative silence of the public on issues of food health and safety. The direct interventions of US sugar industry executives to prevent the WHO's demands for education around the need for a dramatic reduction in sugar consumption and the labeling of food products to assist in this process, is just one example of the routine authoritarian yet perfectly legal behavior of corporate decision-makers (Boseley 2003).

It is important to acknowledge this sort of manipulation in countries such as the United States that have long pointed the finger at nations such as China and Russia, as well as many countries in the Global South, for their authoritarian and corrupt practices. As the documentaries mentioned above, and our conversations with NGO representatives and activists poignantly and convincingly reveal, perhaps one of the biggest "successes" of capitalism has been its ability to allow what can legitimately be seen as corruption to be deemed legal as it runs rampant throughout the system.

This legalized corruption is achieved through the legal mechanisms of elite nepotism disguised as "merit"; lobbying, often through "front groups"; campaign financing; direct bribery through defunding and demonizing of whistle-blowing organizations and groups; ag-gag laws; the revolving door of corporate executives and government representatives; domination of messaging through corporate advertising, and penetration into educational institutions (de Jong & Huluba 2020; Chetty et al. 2016). In addition, as US legal scholar Andrew Spalding points out, the idea of freedom from corruption as a human right, while as old as social contract theory, is new to federal and international law. This means that flagrant violations of human rights by corporate actors go unpunished on a daily basis (Spalding 2014: 1427–8).

It is important to note that our point here is not to suggest that all individuals within the various industries implicated, who help to ensure that the system continues, are selfish, profit-seeking authoritarians; that would be absurd. Rather, it is to point out that many workers and a majority of consumers are blind to the realities of how much their labor and sometimes their life choices are controlled and manipulated through the education system and through powerful lobbies and advertising. Having said this, it is important to acknowledge that many who work in particular jobs may not in fact have had a choice in the matter due to their economic circumstances.

Until recently, the benefits of this system built on exploitation have been enjoyed by the up-until-now growing middle classes of the "advanced" capitalist societies and the expanding middle classes of countries such as China and India. As Noam Chomsky argues through his theory of "manufacturing consent," getting people to buy in to the existing paradigm is not a conspiracy (Herman & Chomsky 2002). Rather, it is a system controlled by capitalist elites seeking to maximize their profits. That the interests of these elites often coincide provides them with great influence over the mechanisms (i.e., media and government) that influence the way people view the world.

Consequently, it is not an organized conspiracy, but rather a confluence of the interests of powerful economic elites who seek the same goals: a system that prioritizes their profit-making activities. So not only do our legal and political systems across the Global North further the interests of these economic elites, but they also reinforce and justify them, and these efforts are complemented through the activities of the education process and corporate-controlled media systems.

The existing system brings with it inevitable harms, the so-called "externalities" that we speak about throughout this book, which manifest as pollution, climate disruption, exploitation of human labor, cruelty to other animals, and so forth. "Adherents" to the system, in the form of laborers and consumers, are often not consciously choosing to harm humans or other living beings. Many of the adherents are average people who simply need a job, who engage in an industry at the lower levels and/or who have been convinced through powerful socialization that what they are doing is okay.

This perpetuation of harm through daily adherence has been a feature of all exploitative processes and systems historically, from racism and patriarchy to homophobia and speciesism. As Nibert notes, "the violence done against the other animals, and the Indigenous Peoples, and the workers, and the consumers, it's all tied together very tightly and it's spiraling out of control" (Interview with David Nibert 2022).

As has been well documented and researched, this broadly accepted "disappearing" or invisibilizing of harm provides a breeding ground for untruths and half-truths in corporate advertising, particularly when it comes to food products and sustainability (Gibbs 2017; Hannan 2020). Some researchers have pointed to "cognitive dissonance" as a widespread phenomenon: the process whereby consumers know that half-truths, and even lies, are being told but still choose to hold onto the belief that the companies do care at some level about the environment (de Jong & Huluba 2020). In their article "The business of lying," Kaylene Williams and co-authors point out: "Even though most people do not like to lie and are basically honest, lying in the business environment still is pervasive. The explanation supposedly is that since fiascos like Enron and WorldCom, businesses are focusing more on compliance with the law alone rather than building cultures where lying is not tolerated" (Williams et al. 2009).

Lies and half-truths can move across the spectrum from exaggerating the benefits of a product to cover-ups of gross violations of human rights and environmental destruction. In a report commissioned by the UN High Commissioner for Human Rights on corporate violations of human rights specifically, business and human rights lawyer Jennifer Zerk points out: "In short, the present system of domestic law remedies is patchy, unpredictable, often ineffective and fragile. It is failing victims who are unable in many cases to access effective remedies for the abuses they have suffered (Zerk 2012).

Even those benefiting most from the existing system recognize the harm being caused by their own wealth-generating activities. In his recent book *Survival of the Richest*, technology expert Douglas Rushkoff speaks about his conversations with some of the world's richest men. He writes:

> The billionaires who called me out to the desert to evaluate their bunker strategies are not the victors of the economic game so much as the victims of its perversely limited rules. More than anything, they have succumbed to a mindset where "winning" means earning enough money to insulate themselves from the damage they are creating by earning money in that way. It's as if they want to build a car that goes fast enough to escape from its own exhaust. (Rushkoff 2022: 10)

Kimberlé Crenshaw, in her work on intersectionality, notes that we can't solve a problem until we have named it, and to name it requires making it visible (Crenshaw 1989). Irrespective of one's political colors, the issues of transparency and accountability should be of concern to any citizen who claims to believe in democracy. When it comes to food, Nibert and others have demonstrated how the growth and profit making of the pharmaceutical industry has depended upon responsive versus preventative health-care systems and a lack of accountability to the public with regard to the health issues associated with our current diets.

If all the information about the health, environmental, and equity issues of our current system were made visible to everyone, and at that point a majority decided they don't care, or at least that they don't care enough, to change their lifestyles, then we are in a different kind of mess, basically meaning that these harms have been democratically accepted. Irrespective of this rather dire but very real possibility, it is the belief of the authors that the public deserves to know the truth about the system and to understand the implications of their/our individual and collective choices.

Much like the horrific pictures that began to show up on packs of cigarettes, this unpopular truth-telling and facing is essential if we believe in transparency and genuine democracy, and if we have a hope for building better systems for the future. We may decide to say "to hell with it" and to continue smoking or we may find that by the time we decide to quit smoking we will already have cancer.

Researchers such as Dahr Jamail and Jem Bendell suggest that it is likely that we already have the climate change equivalent to cancer, but argue that this should inspire us to be thankful for the beauty of this world and the other beings we share it with, to work actively and compassionately together to face the disruption inevitably coming and to enjoy, as much as possible, the time we do have left here.

But even if we do not follow Jamail and Bendell to their very depressing conclusion, we have some very practical questions to wrestle with in the immediate political realm. In a practical sense it is not just about acknowledging our relationship to the land and the responsibility that comes with that regarding future generations, it is also about acknowledging and exposing the sectors that contribute most to GHG emissions and overall pollution, and targeting these industries immediately and dramatically with policy and regulation.

This is precisely what the UN Secretary General António Guterres recently, and non-apologetically, demanded in the lead-up to COP27. While his focus was on the fossil-fuel industry, an industry that certainly must be targeted, we would like to add to this picture the alarming and disproportionate role of industrial *animal agriculture* as one of the single biggest contributors to the current climate crisis, which includes the fossil fuels that go into supporting *animal agriculture* through the crops grown to feed animals (Weis 2013).

This reality is illuminated very dramatically in research results emerging from a comprehensive scientific study conducted by Oxford University and the Swiss agricultural research institute Agroscope. After analyzing data from close to 40,000 farms and 1,600 processors, packagers, and retailers with 40 major foods in mind, they conclude that massive changes are needed in the current food system. While not advocating the "radical" approach of immediate and drastic global regulation advocated by people such as the UN Secretary General, their data is in itself a strong indication that *livestock* production is in fact at the top of the list of industries to focus on in response to climate change and overall environmental pollution.

They propose "an integrated mitigation framework" with four major priorities: (1) digital monitoring of impact by producers (a strategy already successfully implemented by some producers in the United States and China); (2) setting of environmental targets for producers and providing them with incentives such as credits, tax breaks, and reallocation of subsidies; (3) assessment tools that provide producers with "multiple mitigation and productivity enhancement options," and sharing of information about best producer practices; and (4) communication of impacts through the supply chain from producers to consumers (Poore & Nemecek 2018: 992).

The study notes that, in terms of *animal agriculture* specifically, many countries already have "stringent traceability requirements" in place that could assist in communication with consumers through environmental labeling, taxes, and subsidies, thereby reflecting the true costs of production. They also point out that commodity crops are harder to trace and thus they conclude in that case that mitigation may have to focus on producers (Poore & Nemecek 2018: 992).

In placing results from the data on *animal agriculture* specifically into the context of emissions, it is helpful to look at the research of Helen Harwatt from the Animal Law and Policy Program of Harvard University Law School. Harwatt states that "unabated, the livestock sector could take between 37 and 49 percent of the GHG budget allowable under the 2°C and 1.5°C targets, respectively, by 2030. Inaction in the livestock sector would require substantial GHG reductions, far beyond what are planned or realistic, from other sectors" (Harwatt 2019: 533).

In a recent article in *Science* magazine, Michael A. Clark et al. show through comprehensive data and systems analysis that even if we did eliminate all other fossil-fuel emissions today, those coming from the global food system alone would make it impossible to limit warming to 1.5°C and difficult even to realize the 2°C target. They argue that massive changes are needed to our current food production systems if we are to meet those targets (Clark et al. 2020).

We began this book looking at the issue of colonialism and the history of capitalism, dependent as these intertwining forces have been in a relationship of domination of the land and commodification of nature, human, and nonhuman animals. We have indicated that "decolonization" involves not only acting to change our relationship to colonized peoples, nonhuman animals, and lands, but that it is a mindset and involves a proactive process of ensuring our future survival.

While in contemporary society the logic of capital penetrates all social relations to some degree, unless we are living "off grid," it is important to look at what this logic has meant, particularly in the realm of food, and how we may begin to see our way out of the excessive consumerism that permeates our lives.

Dominion over the Land

As we have noted, it is actually impossible to isolate questions of food security and food justice from the discussion of environmental protection and climate disruption, and how the future survival of all species, including human animals, depends entirely upon the natural capacity of the earth to regenerate itself healthfully. The effects of climate change are already seen to be exacerbating food insecurity issues, therefore freeing up land from large-scale agriculture that produces crops for animal feed could facilitate opportunities to grow crops to feed humans (Harwatt 2019).

The World Food Program describes 2022 as a year of "unprecedented hunger." They note that while many countries around the world are facing widespread hunger, some countries, such as Yemen, Somalia, Burundi, South Sudan, and Syria, face famine and/or close to famine conditions. Concern Worldwide US claims that the "toxic cocktail of conflict, climate change, and the Covid-19 pandemic" has left millions of people vulnerable to hunger and starvation. Meanwhile, UNICEF (the United Nations Children's Fund) reports that consecutive years of low rainfall in the Horn of Africa has created "one of the worst climate related emergencies of the past 40 years." In Djibouti, Ethiopia, Kenya, and Somalia, this situation puts 20 million people, including 10 million children, at risk of severe malnutrition and water-borne disease.

As Aviva Chomsky and other historians of the Global South have noted, colonialism and neocolonialism have shaped the history of land control and ownership and the role of workers and farmers on the land. Neocolonialism in its various forms continues to shape the economic, political, and ecological landscape. For this reason, many advocates for food justice such as Chomsky and Shiva talk about food sovereignty as a part of building self-reliant, democratic, and inclusive communities (Shiva 2016a; Interview with Aviva Chomsky 2022). National sovereignty is also highlighted as key to food security and genuine democracy. As Chomsky explains:

> Well, food sovereignty is key, agroecology, peasant rights, that food is a human right. I've been thinking a lot about sovereignty and how sovereignty really doesn't exist in third world countries because of international financial institutions. So, I say that countries should have the right to make their own decisions, but countries are often run by elites who don't have the interests of the poor in mind. I haven't seen the concept of food

democracy. I'm not sure if that's used, but if we think about the way our food system is going globally, it's hell bent on destroying what's left of peasant subsistence agriculture.

I really like the work of Tim Wise on Africa, looking at peasant agriculture. Small-scale peasant agriculture still provides 70% of the food that's eaten in Africa. And those are the things that a food system should be trying to protect and promote, rather than trying to destroy. And you know, there are some parts of the world where it's pretty much already been destroyed, like the United States, not 100% destroyed, but 90% destroyed and we're working on getting rid of the last 10%. But there are parts of the world where it's still predominant and you know that's what we should be finding ways to foster and preserve and honor. The degrowth literature talks a lot about human flourishing, and how society could be organized so that instead of prioritizing growth, it prioritizes human flourishing. (Interview with Aviva Chomsky 2022)

Variations on this argument have been evident in most of our interviews and in much of the literature that we have explored on creating just food systems. The term "degrowth" grew out of a 1972 debate organized by the *Nouvel Observateur,* in which the French philosopher André Gorz explored the relationship of growth to capitalism and warned of the dangers of green capitalism. He asked: "Is global balance, which is conditional upon non-growth – or even degrowth – of material production, compatible with the survival of the (capitalist) system?" (quoted in Duverger 2023; Marty 2023). The debate followed the publication of the Club of Rome report calling for "zero growth."

Editors of *Degrowth Journal,* an interdisciplinary, peer-reviewed, open access publication, note on their website the irony of trying to publish articles that critique capitalism in profit-based journals that hold an "oligopoly on peer reviewed knowledge" (*The Degrowth Journal,* online). In an attempt to bridge the academic/activist gap, they note further:

> Profit making should have no role in science, especially when that very science is about how to escape the social-ecological dead-end created by capitalism. This is not a call for heroic sacrifice; rather, we, the founding editors of this journal, are convinced that by taking a collective stance, critical scholars have the power to change academic culture for the better. (*The Degrowth Journal,* online)

Irrespective of one's views of the capitalist system overall, and acknowledging some of the benefits it has brought to humankind, the point for us here is more about democratic access to knowledge and information. We will return to the concept of degrowth in the final chapter.

When we move away from the practices of commodification, growth, extractivism, monoculture production, factory farming – all inherent aspects of capitalist, fossil fuel-dependent economies – opportunities for building self-reliant and flourishing societies emerge. When we realize and recognize the human-created systems of structural violence that keep these systems going, we begin to see our way out, the possibilities of new paths. Humans have created the messes and, therefore, theoretically they can also fix them.

While humans have suffered from poverty and starvation at various points in history, we can identify structural violence as present when these outcomes are human caused (Leech 2012). To put it bluntly, human beings are indirectly causing many of the recent droughts, fires, and hurricanes that are leading to food insecurity and starvation around the world. In addition, it is important politically to note that it is often not the humans who are suffering who caused the crisis.

This reality has been emphasized by the UN Secretary General António Guterres who helped to put the global fund for "loss and damage" at the top of the agenda at COP27. And while the idea for this fund was accepted by a majority of states, and we could read this as a historic victory of sorts, the lack of movement on what is causing the crisis was not really addressed in any meaningful way. The focus on "aid" and "mitigation" efforts (that generally are costs to taxpayers) is clearly much more politically palatable than changing the rules of the game that allow corporations to continue to make their profits and consumers to buy their cheap goods.

The Violence of a Hamburger and Chicken Nuggets ...

The Animal-Industrial Complex, intimately connected to the history of colonialism and contemporary practices of neocolonialism, has implications for many aspects of life – diets and food as a race and class issue, animal *welfare*, worker safety, and consumer health (Noske 1997; Twine 2012; Nibert 2013). What is apparent is that the various oppressions it facilitates are connected and this also holds true when we begin to examine the implications of the A-IC for environmental degradation and climate change (Twine 2013; Nibert 2017). Mainstream scientists and other researchers have pointed to the fact that *animal agriculture* is one of the largest single contributors to GHG emissions (Koneswaran & Nierenberg 2008; Goodland & Anhang 2009; Harwatt 2019; Lynch et al. 2021). But environmental implications go even further, reinforcing the important connections between *animal agriculture* and habitat and species loss.

But it is not only the industrial production of animals that impacts climate and species diversity. Land degradation and environmental pollution from human settlements and extractive industries have led to unprecedented loss of species from the largest apex predators down to the tiniest insects, including the absolutely vital pollinators who ensure the future of our food supply (Olsson et al. 2022). As we are reminded by scientists, in order to fully grasp the implications of our industrial model of production and consumption on the environment and other animals, we have to move beyond looking at CO_2 emissions alone.

While acknowledging the complexities of comparative analysis of emitting sectors, we must understand the significant role of nitrous oxide and methane from the agricultural sector and include in our framework other significant "planetary boundaries" (Lynch et al. 2021). As Aviva Chomsky notes, in her recent book on climate change, these include: "ocean acidification, stratospheric ozone depletion, biogeochemical nitrogen (nitrogen released by agriculture and industrial processes), phosphorus cycle (phosphorus released into the oceans), global freshwater use, land system use (lands deforested and put into agricultural and urban use), biological diversity loss (species extinctions), chemical pollution, and atmospheric aerosol loading (air pollution, such as particles of dust and soot in the air we breathe)" (Chomsky 2022: xv).

In terms of food production, some analysts have argued that substitutes for products made from nonhuman animals are not helping the situation as they also contribute to deforestation. But it is important to understand that most of the production of crops seen as *meat* alternatives is used to feed nonhuman animals in the *food* system. As pointed out by Ritchie and Roser in *Our World in Data*:

> More than three-quarters (77%) of global soy is fed to livestock for meat and dairy production. Most of the rest is used for biofuels, industry or vegetable oils. Just 7% of soy is used directly for human food products such as tofu, soy milk, edamame beans, and tempeh. The idea that foods often promoted as substitutes for meat and dairy – such as tofu and soy milk – are driving deforestation is a common misconception. (Ritchie & Roser 2021)

While the biggest single driver of Amazon rainforest destruction is the clearing of land for *beef cattle*, some analysts argue that soy production is also an important factor. But Ritchie and Roser highlight the fact that the United States and Brazil together are responsible for nearly 70 percent of global soy production, which is linked to the fact that global *meat* production has tripled in the past half century (Ritchie & Roser 2017a; 2017b; 2019a; 2019b; 2021).

Based on the fact that most deforestation is a result of expanding land use for *beef cattle* and for soy production to feed chickens and pigs, Ritchie and Roser conclude that consumers reducing *meat* consumption is one of the most effective ways to make a difference. They note that "the dominant driver of deforestation in the Brazilian Amazon was the expansion of pasture for beef production. If we look at forest loss from commercial crops – which is mainly soybeans – we see a significant decline, especially following the introduction of 'Brazil's Soy Moratorium'" (Ritchie & Roser 2021).

On a similar note, the research of Western University Geography Professor Tony Weis, whose work focuses on industrial agriculture and political ecology, points to "the destructive connection between agrarian and dietary change and why reducing animal production and consumption must be a key object of environmental and food policy and activism" (2018: 140). In terms of the giant food producers and governments, Ritchie and Roser emphasize zero-deforestation policies focused beyond the Amazon now that production outside that area is rapidly increasing.

While citizens in the Global North may appreciate the Amazon rainforest in some distant sense, and with a vague awareness of the importance of rainforests as carbon sinks, likely they do not fully grasp our dependence on the Global South in general, and the Amazon in particular, for everything from our pharmaceutical products to our junk food diets. Scientists in the Global South such as Rita Mesquita (interviewed by journalist Dahr Jamail for his book *The End of Ice*), explain how feeding Western diets is one of the primary drivers of the devastation and violence taking place in countries such as Brazil (Jamail 2020).

Clear cutting for timber and *livestock* farming is at the root of the problem. As Jamail explains, "Rates of deforestation across the Amazon are increasing, and Brazil has the highest rate of assassinations related to environmental and land agrarian reform globally. By

2016, activists were being killed at a rate of nearly four people every single week worldwide. Brazil saw the highest rates, with forty-nine killings, many of them in the Amazon" (2020: 170). Meanwhile, Mesquita states: "What Brazil is doing in the name of mining and industrial agriculture is mind-bending, we are trashing our protected areas" (quoted in Jamail 2020: 171). Scientists from Brazil highlight the massive international pressure on the Brazilian government to clear cut, identifying deforestation for cattle ranches selling beef to North America and Europe (Jamail 2020: 177).

In addition to forests, *animal agriculture* has huge implications for our freshwater resources. Out of concern for these pressures, organizations such as UNESCO have been reporting on the "water footprint" of the *meat* and *dairy* industries for many years, noting this blind spot in policymaking on food security and environment. Back in 2010, the UN agency stated:

> Managing the demand for animal products by promoting a dietary shift away from a meat-rich diet will be an inevitable component in the environmental policy of governments. In countries where the consumption of animal products is still quickly rising, one should critically look how this growing demand can be moderated. On the production side, it would be wise to include freshwater implications in the development of animal farming policies, which means that particularly feed composition, feed water requirements and feed origin need to receive attention. (UNESCO 2010: 39)

Statistical documentation over many years reveals that products made from nonhuman animals have a far larger water footprint than crop products, including when measured per calorie. *Beef* alone requires 20 times more water than root vegetables and cereals. In terms of plant-based dietary protein sources, pulses require six times less water than *beef* (Hoekstra 2012).

The Water Footprint Network (WFN) argues that in terms of freshwater, it is clearly more efficient to obtain our proteins, fats, and calories from crops rather than products made from animals (Water Footprint Network n.d.). But, given the profit interests of the *meat* and *dairy* industries, it is not surprising that they have shown little interest in acknowledging or properly monitoring their water usage. As the WFN notes, these industries account for more than a quarter of the global water footprint of humanity. Governments around the world have failed to seriously address this issue:

from the governmental side, hardly any attention is given to the relationship between animal products and water resources. Nowhere in the world does a national water plan exist that addresses the issue that meat and dairy are among the most water-intensive consumer products, let alone that national water policies somehow involve consumers or the meat and dairy industry in this respect. Water policies are often focused on sustainable production, but they seldom address sustainable consumption. (Hoekstra 2012: 3)

So, while the silence of corporations benefiting from this reality may provoke anger, the lack of action by elected governments is also shameful and only adds more fuel to the argument that our elected representatives are more interested in protecting industry, gross domestic product, and economic growth than they are in protecting citizens or confronting climate change. Given this reality, it is obvious why the protection and well-being of other animals as a concern is not even close to penetrating national agendas.

If one's starting goal is the need to continually generate and expand profit, then the A-IC may be one of the most "efficient" systems we could conceive of. This is in large part because the true costs of this efficiency are socialized while the profits are privatized. However, if the starting goal is the need to create a sustainable and just food system, then the A-IC is not only horrendously inefficient, but its continued existence also guarantees failure.

Our diets are also contributing to increased health-care costs due in part to the epidemic of obesity, increasing cases of heart disease, and diabetes, including increasingly early onset. One of the most rigorous and best-selling books on health and nutrition, *The China Study*, which *The New York Times* names as the "Grand Prix of epidemiology," reveals that these "diseases of affluence" are now prevalent in societies such as India and China, whose growing middle classes have adopted Western dietary practices (Campbell & Campbell 2006).

Recovering from the Green Revolution

The so-called Green Revolution was seen by its most vigorous proponents as an opportunity to use science to feed the world by utilizing land and inputs more efficiently to increase yields. The growth and power of the A-IC paralleled and complemented the general industrialization of food production after World War Two and should be analyzed and evaluated as part of an overall shift in methods related to the Cold War context and the economic priorities of capitalist food producers (Holt-Giménez 2017a). The Green Revolution, emphasizing high-yield varieties of agricultural products (particularly wheat, rice, and corn), hybridization techniques, and heavy reliance on fossil fuel-based pesticides and fertilizers, was promoted around the world and particularly in the Global South.

Beginning in the 1950s and motivated by fear of communist and socialist revolutions, the Rockefeller and Ford Foundations along with the World Bank encouraged Global South countries to move in this direction. Vandana Shiva and others (Shiva 2016a; 2016d; 2022; Holt-Giménez 2017a) have monitored and critiqued the development of this chemical-based agricultural model forced on beleaguered, indebted, and colonized countries of the Global South, serving as what she calls "an antidote to social change" while deepening inequality (Interview with Vandana Shiva 2023).

Smaller-scale farmers were forced to leave their lands as they were unable to afford the credit payments for inputs associated with this mode of farming. Other impacts included desertification associated with the diversion of vast quantities of water needed in production, and significant and ongoing decreases in agricultural diversity. It is a dubious starting point that some of the chemicals used to fuel the Green Revolution such as pesticides, that are now mainstays in global food production systems, were derived from military defense technologies to create "nerve gas."

The use of pesticides and fertilizers is a requirement in the large-scale monoculture production process. While Shiva notes that the production of rice and wheat did increase, this does not mean that there has been an absolute increase in the production of food. Quite the opposite in fact; there has been a significant decline in the production of vegetables and legumes. For some, science can be seen as the problem, as part of the colonial history of capitalism, with the solution being a truly green revolution. As Aviva Chomsky points out, "I feel like that's the issue, to disentangle science from its terrible social and cultural history, how capitalism and imperialism have shaped what we think of as science" (Interview with Aviva Chomsky 2022).

As Shiva noted at a 2022 Seed Fair in Florence, Italy: "There are two trends. One: a trend of diversity, democracy, freedom, joy, culture – people celebrating their lives ... And the other: monocultures, deadness. Everyone depressed. Everyone on Prozac. More and more young people unemployed. We don't want that world of death." She continued by pointing out, "We would have no hunger in the world if the seed was in the hands of the farmers and gardeners and the land was in the hands of the farmers. They want to take that away."

But Shiva is not content to just protest and critique the system; she has worked to promote and create concrete alternatives in India, including establishing a biodiversity farm in the northern state of Uttarakand. According to Shiva, "Navdanya is an Earth-centric, women-centric and farmer-led movement for the protection of biological and cultural diversity. We live and practise the philosophy of Earth Democracy as one Earth Family (Vasudhaiva Kutumbakam) with no separations between nature and humans and no hierarchies between species, culture, gender, race, and faiths" (Shiva 2022).

A key part of the movement has been seed saving, seen as an act and movement of resistance against corporate monopoly monoculture production. Navdanya has created seed banks in 22 states, ensuring that future generations will have access to healthy diverse food options because these seeds are available for free to farmers in contrast to the patented seeds produced by corporations such as Monsanto and Cargill. Inspired by the Chipko movement, a non-violent women-led movement in the 1970s aimed at protecting India's forests from clear cutting, Shiva herself is one of the original "tree huggers."

Underscoring the reality of colonialism and climate change's non-jurisdictional nature, the case of India, whose experience is repeated in other countries in the Global South, makes clear the suffering that comes as a result of a fossil-fuel industrialism, which has not benefited a majority of Indian people. The Himalayan glaciers lose one and a half feet of snow every year as a direct impact of fossil-fuel industrialism (Maurer et al. 2019). This depletion directly impacts farming communities and access to water in general. Similar stories of climate disruption exist elsewhere, with consecutive years of droughts and flooding in countries such as Somalia, Ethiopia, South Sudan, and Yemen decimating crops and creating massive food insecurity and famines unprecedented in the past century according to the United Nations.

The agonizing stories and images from these countries, received daily by the authors in "urgent action" requests, brings the acute need for compassionate food systems painfully to light. But, unlike "acts of God" that we can feel empathy and sadness about, and provide some aid towards, these climatic events have direct links to the rampant fossil-fuel production and consumption fueling Global North lifestyles.

As noted above, when we simply measure emissions within national boundaries, we do not account for who is buying the products being made within those boundaries. One look at many of our product and clothing labels makes clear that our patterns of consumption in the Global North are massively implicated in the high levels of emissions in countries such as China and India, and are fueling climate disruption around the world.

The Plant-Based Treaty: Moving Toward a Global Shift in Culture?

We have noted that a shift in culture is at the root of the move towards plant-based diets. Many organizations worldwide are currently working in this area: the Vegan Society (UK), PETA (US), Food Empowerment Project (US), Animal Aid (UK), Animal Save Movement (worldwide), Proveg International, and many others. In addition, we see policy-oriented grassroots initiatives such as the call for a "Plant-Based Treaty" to accompany the UNFCCC/Paris Agreement with the goal "to put food systems at the forefront of combating the climate crisis." The Plant-Based Treaty points to the connection between intensive *animal agriculture* and the rapid degradation of critical ecosystems with a call for moves to sustainable plant-based diets.

What is now, finally, becoming part of a more mainstream discourse is the fact that while all humans in some way affect ecosystems, anthropocentric climate disruption and crisis is not ultimately caused by all humans equally. Why is this important? As the UN COP27 conference made clear, the rich countries of the Global North have disproportionate responsibility to solve the problem and to provide "cost and damage" to those nations facing the worst consequences simply due to the fact that they have been the key "perpetrators" and the key beneficiaries. In a global system that is, at least theoretically, rooted in the rule of law, it is imperative that justice is served.

The fear of grassroots movements around the world is that "voluntary" contributions to global funds and voluntary moves towards "phasing down" rather than phasing out fossil-fuel dependence by governments cannot solve the crisis in time to ensure a livable planet for future generations. The sad reality is that COP27 ended with no agreement on fossil-fuel phase-out for Global North countries. This left many in the Global South frustrated that the obvious connection between "loss and damage" and keeping temperature increase to the 1.5 degree mark are not being recognized as inextricably linked.

As Aviva Chomsky and others have noted, at least part of the problem is political in terms of the way we calculate emissions by country, a method that does not reveal accurately which countries are most responsible. Rather than calculating total CO_2 in the atmosphere – the cause of climate change – the international community measures CO_2 emissions per country per year, which is why China is constantly blamed for being the worst offender. However, counting per capita and cumulative emissions, on the other hand, makes clear that the European Union and the United States have "by far the greatest historical responsibility" and that their citizens are also emitting the most CO_2 through consumption (Chomsky 2022: 113). In other words, a significant portion of China's CO_2 emissions are generated producing goods for consumers in the Global North.

Many argue that, in addition to confronting overconsumption, handling inequality and poverty is a critical part of the solution to climate change. As Aviva Chomsky notes:

> A corollary to our capacity to imagine plenitude in a lower consumption lifestyle is a robust social safety net that takes basic needs and rights out of the insecurity of the market. If we knew we could rely on free daycare and college, real national healthcare and pension systems, and the security that our basic needs like food and housing were considered human rights and guaranteed by the public sector, it would be a lot easier to contemplate working fewer hours and redefining a quality of life that was not based on ever-increasing consumption. (Chomsky 2022: 98)

The shift to more sustainability overall goes hand-in-hand with creating social safety nets so that contemplating working and consuming less makes sense (Schor 2011). It also requires addressing the inequalities and exploitation experienced by both human and nonhuman animals in the existing industrial food production system.

3 Working in Hell

Labor in the Industrial Production of Animals as Food

> [The pigs] were brought in knowing, essentially knowing what was about to happen to them. So, their sense of panic was the first thing that you noticed. It was the same with all three [sheep, cows, and pigs]. I think they essentially understood what was going on at that point.
> – Paul Rooke

When considering what food in a just society might look like, recognizing the needs of both human and nonhuman animals becomes an essential component of challenging inequality and must factor prominently into any proposed changes to the Animal-Industrial Complex. As we've discussed in previous chapters, capitalism, with its commodification of land, labor, and nonhuman animals, "has not given us an equitable, healthy, and resilient food system" (Holt-Giménez 2017a: 70).

While the focus of this chapter is on the labor utilized in the industrial production of nonhuman animals as *food*, it is important to begin with the acknowledgment that agricultural work in general is "among the most dangerous, undervalued, and precarious forms of work. Additionally, it is done in geographically isolated areas, which poses particular problems in terms of transparency and availability of redress for workers (and working animals!)" (Blattner et al. 2021: 260).

The logic of capital means that workers in food production routinely face low wages and, in some cases, unsafe working conditions. This includes workers who pick fruits and vegetables as well as those in fast-food restaurants and grocery store retailing. Agricultural work often relies on temporary foreign and undocumented migrant workers who face additional precariousness because they may rely on their employers for necessities such as housing and access to benefits, and do not enjoy the same rights as other citizens.

However, workers in the industrial production of nonhuman animals as *food* often face additional, and unique, challenges. A key question that many may find themselves asking is: in what ways is this type of work different from other manufacturing or industrialized work? In addition, we also consider the literature on the division of labor in industrial production of animals as *food*, a process that helps workplaces compartmentalize operations so that some workers may be removed from certain work – as some working in slaughterhouses, for instance, are often physically removed from the kill floor. In what are commonly referred to by industry as *meatpacking* and *chicken processing*, nonhuman animals enter the slaughterhouse alive, are killed, dismembered, and leave as *food*.

Some industrial animal-centered work, such as happens in slaughterhouses, is often designated as dirty work, as it involves exposure to blood, feces, and routinized violence, conditions which can lead to externalities or spillover into individual workers' lives, and into their communities. In addition to considering the labor of human workers, we also address issues raised by the most recent research on the "labor" of other animals. Until recently, little attention has been paid to the labor of nonhuman animals in the creation of *food*, so it is important to consider what this entails and how people portray the production of products made from nonhuman animals such as *meat*, *dairy*, *eggs*, and *fishes*, as natural and inevitable.

Human workers and consumers engage in a type of boundary work to maintain a sense of separation from other animals constructed as *food*, which is made easier with a division of labor for some workers, and lack of transparency for consumers. The issue of transparency is of particular importance because opacity insulates consumers from the worst practices of food production, and it is important to make clear why transparency is essential in the building of a democratic and just food system. As we have said previously, transparency helps citizen-consumers make informed decisions about what to consume, and also to demand political and economic change when necessary.

Here we also consider the important and empowering concepts of resistance and solidarity. Some workers challenge unsafe and unjust working conditions by forming or joining labor unions, encouraging citizen-consumer boycotts, and routinely rejecting employment in slaughterhouses, for instance. Nonhuman animals may resist by running away or refusing to be compliant in the harms perpetrated against them.

Labor in the Animal-Industrial Complex, COVID-19, and the Logic of Capital

In 2021, an article published by the Canadian Broadcasting Corporation (CBC) focused on the deaths of several slaughterhouse workers in Alberta from COVID-19 (Dryden & Rieger 2021). The authors were quick to highlight that the workers they spoke to said that their workplace problems pre-dated the pandemic. The workers pointed out that from the beginning of their employment, they had experienced pity for the animals, as well as enduring the horrible odor of blood and feces on the kill floor and feeling like a machine.

One worker described the fast pace of the *production* line (45,000 pigs are killed and *processed* at this plant each week), and the repetitive motions that can lead to bodily deterioration as well as serious injury. News reports indicate that the injury rate at Olymel in Red Deer, Alberta, was an alarming 18.1 per 100 full-time employees, and this was considered high even within an industry known to be dangerous (Dryden & Rieger 2021).

At this same slaughterhouse, the company decided to increase the line speed during the COVID-19 pandemic; increasing the number of pigs being slaughtered and *processed* each day from almost 7,000 to 10,000 (Freeman 2021). This occurred within the same physical dimensions of the plant, with more workers *processing* more animals (Freeman 2021). This increase meant more overcrowding as well as the potential for more stress and anxiety for both the workers and the nonhuman animals. Highly mechanized industries such as slaughterhouses try to keep costs low by speeding up production lines and, as a result, workers end up having to adapt to the changes to the production line speed rather than being taken into consideration in the implementation of any such changes (Dryden & Rieger 2021).

It is also important to note, however, that the conditions described at this plant are not unique, and to recognize the ways in which the pandemic, and institutional responses to it, increased the threat of illness for slaughterhouse workers, their families, and their communities. For instance, a recent article in the *Proceedings of the National Academy of Sciences* examined the relationship between slaughter and *meatpacking* facilities and COVID-19 transmission in the United States. It estimated that such plants were "associated with 236,000 to 310,000 COVID-19 cases (6 to 8% of total) and 4,300 to 5,200 deaths (3 to 4% of total) as of July 21 [2020]" (Taylor et al. 2020: 31706). At a community level, those that contained slaughter facilities suffered higher rates of COVID-19 infection (Taylor et al. 2020) as well as ongoing economic hardships due to low wages, which facilitate the need for many workers to have a second or third job (Broadway 2013; Dryden & Rieger 2021).

The COVID-19 pandemic, and the ensuing reporting on the *depopulation* in the industrial production of other animals as *food*, brought attention to their commodification and eventual death, and at the same time revealed the experiences of workers. A news article entitled "Piglets aborted, chickens gassed as pandemic slams meat sector" described what happened when COVID-19 outbreaks forced closures of slaughter facilities during the first wave of the pandemic. When plants closed it created a backlog that meant farmers had *market-ready* animals that were now staying on their farms after they were the required size for slaughter. These nonhuman animals obviously needed to continue to eat, and as they grew larger, crowding ensued and costs to the farmers rose (Polansek & Huffstutter 2020). Many farmers started the process, referred to as *depopulation*, of killing nonhuman animals *en masse* to alleviate crowding, make room for new animals and/or to save money (Kevany 2020a; 2020b; Baysinger & Kogan 2022; Reyes-Illg et al. 2022).

Farming is a profession that has enormous challenges and stress. Farmers deal in issues of life and death every day, whether growing crops or raising other animals, as well as facing economic pressures and hardships. We asked medical anthropologist and farmer Joe Parish about how farmers cope with the stresses associated with this profession, and he replied, "Not well in some cases – it has one of the highest suicide rates as a profession. There are actually hotlines for farmers who are experiencing depression ... I know in the prairie provinces they have provincial hotlines because people are not only dealing with death, [but] they're also dealing with economic hardship, right? They're struggling economically" (Interview with Joe Parish 2023). According to the US Centers for Disease Control and Prevention, farmers in the United States commit suicide at twice the rate of people in most other occupations (Forrest 2022; Miller & Rudolphi 2022).

During the on-farm *depopulations* that occurred in the first wave of COVID-19, farmers and veterinarians reportedly suffered the ill-effects of having to kill the nonhuman animals they were raising, or those in their care. All this was happening as the slaughterhouse industry experienced sporadic closures and reduced capacity due to worker shortages. These industry changes resulted from both COVID-19 outbreaks and supply chain reductions in the demand for certain animal-based *products*, such as *pork*, in food service industries (Labchuk 2020; Baysinger & Kogan 2022; Reyes-Illg et al. 2022). *Euthanasia* (defined as *humane* killing) of other animals is known to be stressful for human workers, such as veterinarians and farmers. Research suggests that such workers can experience a range of mental health symptoms such as depression and may cope by using alcohol or drugs (Shearer 2018).

Many researchers have examined the "caring-killing paradox," first developed by sociologist Arnold Arluke, and found it to be a helpful tool to think through how workers come to terms with having contradictory responsibilities for the nonhuman animals with which they interact (Arluke 1994; Arluke & Sanders 1996; Shearer 2018; Tallberg & Jordan 2021; Baysinger & Kogan 2022). On the one hand, they are responsible for caring for other animals (i.e., in animal shelters, at farms, as veterinarians), and yet, on the other hand, they are sometimes responsible for killing those same nonhuman animals (Arluke 1994; Arluke & Sanders 1996; Shearer 2018; Tallberg & Jordan 2021; Baysinger & Kogan 2022). This paradox can lead to significant psychological stress. In his study of US *beef* producers, sociologist Coulter Ellis found that farmers take part in what he calls "boundary labour" to help explain the ways that they can manage the inconsistencies involved in treating other animals as individuals on the one hand, but also as "commodities" on the other (Ellis 2014: 111). Sociologist Rhoda Wilkie argues that in the commercial production of *livestock*, workers are supposed to be both "detached" but still provide "care" to the nonhuman animals in their charge. Wilkie frames this discrepancy as "concerned detachment" (2005: 218).

Evidence suggests that on-farm *depopulation* sometimes relies on forms of mass killing that are not always quick or pain free (Reyes-Illg et al. 2022; Whiting & Keane 2022), and that it is important for workers' mental well-being that the methods used be *humane* (Shearer 2018). One of the methods reportedly used in the United States to kill pigs during the pandemic included ventilation shutdown and steam heat (VSD+). This is a process whereby the heat is turned up and hot air is blown onto the animals, causing them to die of heat stroke (Whiting & Keane 2022: 861). Heat stroke causes a variety of painful symptoms prior to death: vomiting, diarrhea, brain injury, and shock, among others (Reyes-Illg et al. 2022). Serious questions and concerns have been raised about how quickly pigs die under such methods, and the amount of suffering they experience. From an animal *welfare* perspective, this form of killing is known to cause prolonged suffering, so it should not be considered a *humane* method of killing because it is cruel (Whiting & Keane 2022: 861).

Under the Canadian *Code of Practice for Care and Handling of Pigs*, a conditional method of *euthanizing* piglets is the use of blunt force trauma. It is described in the *Code* as "grasping the hind legs of the piglet and striking the top of the cranium firmly and deliberately against a flat, hard surface" (NFACC 2014: 61). While this method of killing young pigs is permitted, the code notes that "alternative methods should be actively considered to ensure that criteria for euthanasia can be consistently met" (NFACC 2014: 61). One can easily imagine how such a method of killing would take an emotional toll on farm workers, farmers, or veterinarians.

In addition to the essential consideration of nonhuman animals' suffering under such conditions of death, it is important to consider how killing a large number of healthy animals and/or the utilization of less *humane* killing methods may increase psychological distress in veterinarians participating in the killing. This distress may manifest as trauma and burn-out (Reyes-Illg et al. 2022). Killing nonhuman animals seems to take a toll on workers, such as farmers and veterinarians, as "[m]ental illness characterized by depression and anxiety is prevalent for those required to participate in the euthanasia of animals irrespective of their background or level of preparation. Euthanasia is difficult and the greatest cost to humans may not be measurable in dollars and cents" (Shearer 2018).

There are important structural conditions that make the possibility of mass, on-farm *depopulations* more likely, the number of lives lost much higher, and the choice of killing method potentially less *humane* (Reyes-Illg et al. 2022). Many of the changes we will examine, in relation to larger facilities and corporate concentration in the slaughter industry, are also relevant to farming as well. Farm operations have grown larger, and large *farms* account for a higher percentage of the overall production of nonhuman animals being raised for consumption. For example, in the period from 1997 to 2017, the number of US farms that raised *hogs* declined by about half, but the remaining *farms* grew larger so that operations with more than

5,000 animals accounted for 72.8% of pig production by 2017 (Davis et al. 2022: 5–6).

In Canada, farm sizes have increased as well. By 2022, the average *hog* operation had a little over 1,900 animals (Statistics Canada 2023). Some provinces had substantially larger operations, such as Manitoba's average of 4,831 animals per operation (Brisson 2014: 9). Both corporate concentration in processing and increased size of animal-based operations are factors that can help us to understand why there is less margin for error when a slaughter facility shuts down for a period of time (Reyes-Illg et al. 2022).

Corporate concentration in the slaughter industry means that if one plant closes, there may be no other location for farmers to access. When *depopulation* is facilitated, because of slaughterhouse shutdowns or the threat of severe disease outbreaks – the larger the farm, the larger the number of animals that it will be deemed necessary to kill (Reyes-Illg et al. 2022). And unlike small-scale farms, decisions become much costlier and consequently may be made regarding the economics of the situation rather than simply on the basis of health considerations (Porcher 2011: 7).

During the pandemic, in Canada and the United States, it is estimated that millions of nonhuman animals were killed on-farm due to slaughterhouse shutdowns (Polansek & Huffstutter 2020). Guidelines for on-farm killing of nonhuman animals under emergency situations specify that the methods used should be as *humane* as possible, and that they be species appropriate. For example, the *Canadian Code of Practice* for *poultry* stipulates that "the methods employed for destroying large numbers of birds in emergency situations need to be as humane as possible given the circumstances" (NFACC 2017: 59). Species-specific guidelines for on-farm *depopulation* under emergency situations, such as a disease outbreak or slaughterhouse shutdown, also indicate preferred methods, conditional methods (permitted in constrained circumstances), and those methods that are unacceptable. They also "require" that farmers develop a plan for on-site *euthanasia* and that they do so in consultation with a licensed veterinarian (NFACC 2014: 41; NFACC 2017). Requirements under the *Code* "refer to either a regulatory requirement, or an industry-imposed *expectation* outlining acceptable and unacceptable practices and are fundamental obligations relating to the care of animals" (our emphasis) (NFACC 2023). The NFACC indicates that such requirements may be deemed necessary by industry associations, and that some may be "enforceable under federal and provincal regulation" (NFACC 2023). As we will discuss further in chapter 5, these codes are not legally binding and largely depend upon industry self-policing, but the World Animal Protection organization does note that six provinces now reference these codes in their animal protection regulations (2020).

A variety of factors are taken into consideration in the decision-making process regarding methods for on-farm *depopulation*, including available resources (e.g., economic considerations) or access to the nonhuman animals (e.g., difficulty of individually accessing animals in large production facilities). Such considerations can be used to justify the utilization of less *humane* methods, such as conditional methods discussed above (ASPCA 2020).

Some of the methods used to kill animals in both the United States and Canada, as reported in a variety of news stories during the COVID-19 pandemic slaughterhouse closures, ranged from shooting pigs, blunt-force trauma for piglets, and gassing chickens. For farmers that breed animals for populating other farms, the market for baby animals for farming halted during the pandemic. The market value for piglets in Canada, for instance, fell to zero because of slaughter-house closures in the United States (Polansek & Huffstutter 2020). Farmers in this situation sometimes choose to medically induce abortions in pregnant pigs (Polansek & Huffstutter 2020).

In order to help explain why the pandemic had such dangerous and deadly consequences for slaughterhouse workers and for nonhuman animals raised and processed as *food*, we need a broader historical, political, and economic context. Prior to the 1960s in the United States and 1970s in Canada, large *cattle* and *hog* slaughterhouses were located near city centers (Stull & Broadway 2004; Fitzgerald 2010; 2015). The shift of this work from urban to rural areas was accompanied by other changes, such as the construction of more mechanized and single-species plants (Stull & Broadway 2004: 16–17). And then, in the 1970s, newer companies added changes to the slaughterhouse industry, such as new ways of cutting, increased automation, and faster assembly lines. These companies also made major shifts in their labor strategies, including the elimination of collective bargaining in some areas and the active recruitment of an immigrant workforce (Fitzgerald 2015: 49).

While there has been a marked decrease in the income and collective bargaining power of slaughterhouse workers since the

1960s, it became most pronounced in the 1980s, as a result of an overall decline in union influence as part of wide-reaching neoliberal policy shifts (Haedicke 2013: 127–8). All of these industry changes have helped to ensure a workforce that is easily replaceable and generally more compliant.

When we asked professor and public sociologist Michael Haedicke about the ways in which slaughterhouse work is shaped by economic and political forces, he highlighted three important factors that undermined the ability of unions to have a powerful presence in this industry in the latter part of the twentieth century:

> Number one is the consolidation of the industry, number two [is] shifts in public policy, and number three [is] the movement of meatpacking activities from urban areas to rural towns in the Midwest where there was not a familiarity with unions. [In those rural areas] there were not a lot of other job opportunities besides meatpacking, and so there was a lot more deference on the part of the town leaders and also workers to firms. And then ... meatpacking firms [hired an] immigrant workforce. (Interview with Michael Haedicke 2022)

Haedicke went on to say that *meatpacking* workers in the United States are often refugees, undocumented workers, and those with temporary status. He also discussed how this is "another liminal status that provides you with some rights as workers but makes it much more difficult for you to vigorously resist the demands of an employer" (Interview with Michael Haedicke 2022). It is essential to note that these changes did not occur by happenstance, as "[t]he movement of meatpacking from the urban areas to rural areas in the United States was actually a deliberate effort to break the power of unions in workplaces as well. So, it didn't just happen, it was part of a concerted effort by industry leaders" (Interview with Michael Haedicke 2022).

Conceptualizing the political and economic influences that facilitated these changes, writer, activist and academic Raj Patel and professor of world history and world ecology Jason Moore explain that cheap *meat* is made by cheap labor. In the United States, they argue, cheap labor for the slaughterhouse sector is largely provided by Latino immigrants (2017: 156). Rarely do popular debates regarding immigration illuminate the conditions that deliberately made slaughterhouse work in the United States *desirable* to Mexican immigrants, such as the shift in union influence and replacing those well-paying jobs with low-paying jobs for immigrant laborers (Patel & Moore 2017: 156–7). Such debates also glossed over the ways in which NAFTA (North American Free Trade Agreement) destabilized farming in Mexico, which created an influx of unemployed workers from Mexico into the United States as a cheap source of labor (Patel & Moore 2017: 157).

> **Michael Haedicke, Professor and Public Sociologist:** One of the major difficulties is that for many consumers today, there is just not a sense, or an understanding, of what is involved in the production of food. I teach a sociology of food course pretty regularly, and on the first day I ask people to think about their most recent meal and to explain how far back they can trace the ingredients of that meal. So, whatever it is, where did it come from? Do [you know] anything about where it came from, not only the geographical location, but also say, do you know anything about the experiences of the people that were part of bringing that food to you as the end consumer, as the eater of that food? And unsurprisingly, with a few exceptions, the students really cannot do that. They say it came from the store, that's as far back as they can trace it ... And from a sociological point of view, really importantly, zero understanding of what the experiences were of people who interacted with the food before it came to the end consumer ...
>
> There's a lack of understanding of how food systems are organized, of what people's experiences are, what goes on in food production. And that is a feature, I think of the complexity and the industrialization of food production, and of the way food is marketed ... in the grocery industry. One of the things that's happened over the past couple of decades is an emphasis on the experience of food shopping, really kind of launched by Whole Foods or brought to the larger market by Whole Foods. And today you see other sorts of grocery chains really trying to make the experience of going to the grocery store more interesting, you know, more engaging, more stimulating for consumers ... but it does make it difficult to sort of look back, right? Because the emphasis, people's focus, is on the grocery store itself, the retail environment, not what happens before

> that environment. So that's what I meant by the complexity that separates producers and consumers.
>
> And then along with that, another layer of division, right, another layer that exists in those systems is the fact that people who are privileged as consumers are typically racially, ethnically, linguistically, in terms of legal status, separated from those who are doing the work of food production. That's not always the case, but it's often the case, that the people, those who run grocery companies think about most frequently, are affluent well-educated white women, because that's who buys the most groceries in the US. And those who are doing the work of food production are not that demographic typically. They are people of color, people with liminal legal status in the United States, people without advanced education, and that's another sort of disconnect or another barrier in food systems. That makes that sort of solidarity more difficult to establish. (Interview with Michael Haedicke 2022)

The Health and Demographics of Workers

From an intersectional perspective, it is important to discuss the fact that this dangerous work also has obvious racial, gender, and class-based dimensions. US *meatpacking* companies compensate for the high turnover by often employing "legal and illegal migrants from Mexico and Central America" (Broadway 2013: 47). Similarly, Canadian companies often rely on recent immigrants, refugees, and temporary foreign workers in the slaughterhouse industry (Broadway 2013: 47).

Of "front-line meat-processing workers in the United States, 45% are categorized as low income, 80% are people of color, and 52% are immigrants, many of whom are undocumented" (Taylor et al. 2020: 31707). In Canada, it is reported that incomes from slaughterhouse employment are rarely enough to survive on, and many employed in this industry depend on secondary employment to make ends meet. At one pig slaughter and *processing* plant in Alberta, three-fifths of the employees worked additional jobs (Dryden & Rieger 2021).

A 2022 report published by the US Congressional Select Subcommittee on the Coronavirus Crisis revealed that "[I]n addition to featuring a heavily concentrated market, the meatpacking industry has a high degree of coordination among competitors, often through participation in powerful trade associations such as the North American Meat Institute (NAMI) and the National Chicken Council (NCC)" (US House of Representatives 2022: 5). On the theme of corporate corruption, lying and unethical behavior, raised in earlier chapters, "[t]hese organizations facilitate coordination amongst their members – some of whom have recently been investigated for price fixing – for matters ranging from public affairs strategies to crisis response, including their response to the coronavirus crisis" (US House of Representatives 2022: 5).

This report maintained that the five US *meatpacking* companies that were the focus of the investigation, and their representatives, knew how severe the risk of COVID-19 was for their employees where production continued as usual, and that they also knew there really was no danger to national food security if they lowered production output or had short-term shutdowns due to outbreaks (2022: 10–11). The report concludes that even knowing these risks, they successfully lobbied government agencies such as the US Department of Agriculture "to keep workers on the job in unsafe conditions, to ensure state and local health authorities were powerless to mandate otherwise, and to be protected against legal liability for the harms that would result" (2022: 32–3).

We can see how Garry Leech's argument concerning structural violence as an outcome of the logic of capital is applicable to the situation in the United States where these actions resulted in thousands of infections, and hundreds of deaths, during the first year of the pandemic (Leech 2012). In fact, at least 59,000 workers from these companies were infected in the first year of the COVID-19 pandemic, and at least 269 workers died. Furthermore, there were spillover infections and deaths to workers' friends, families, and the wider communities in which they lived (US House of Representatives 2022: 33).

The Select Subcommittee investigated five of the largest companies involved in the US *meatpacking* industry: JBS, Tyson, Smithfield, Cargill, and National Beef. Each company's *meatpacking* facilities had large outbreaks of COVID-19 during the initial year of the pandemic (US House of Representatives 2022: 7). The document details: "the meatpacking industry's efforts – aided extensively by Trump's USDA and White House officials – led to policies, guidance, and an executive order that, individually and altogether, forced meatpacking workers to continue working despite health risks and allowed companies to avoid taking precautions to protect workers from the coronavirus, ultimately contributing to thousands of worker infections and hundreds of worker deaths" (US House of Representatives 2022: 7).

At the same time, profits continued to be enormous:

> Tyson reported a net income of approximately $3 billion in 2021, and $2 billion in 2020. JBS reported a net income of approximately $4.2 billion in 2021, and $937 million in 2020. Meatpacking companies' profit margins in recent years have been so high that even industry lobbyists have recognized the possibility of poor optics. In one email obtained by the Select Subcommittee discussing a pandemic-related proposal by USDA to require and subsidize hazard pay for meatpacking workers through corporate tax breaks, a meatpacking lobbyist asked a Tyson lobbyist "Given where margins are, do we want to publicly support a tax break for [meat] packers?" (US House of Representatives 2022: 5)

These facts demonstrate that this is not an issue of corporations being unable to afford to pay their workers appropriately. Rather, it relates clearly to the logic of capital in which companies squeeze out as much value and profit from their workers as possible in order to maximize corporate and shareholder profits. To put this in perspective, "Cargill Ltd. is a Canadian subsidiary of the US-based Cargill, which reported revenue of $113.5 billion US and net earnings of $2.56 billion [in 2019]" (Dryden & Rieger 2020). During 2020, Cargill Inc. "reported net income of US$3 billion ... up 17% from the previous year" (Blas 2020). The 125 family members who own the company received their biggest dividend ever, $1.13 billion (Blas 2020). According to the Collective Agreement, the starting wage in 2020 for the production line at the Cargill *cattle* slaughterhouse in High River, Alberta, was $19.50 per hour (UFCW Local 401 2021).

This corporate concentration of food processing in the United States and Canada poses serious issues for the food supply chain; it also runs counter to a democratic food system because decision-making power regarding what to produce and how to produce it rests in the hands of a very small number of corporations, as do the profits generated. In the United States, corporate concentration in slaughter and *meat packing* means that a dozen plants are responsible for more than half of the *beef* produced, and an additional 12 plants are responsible for over half of the *pork* produced (Taylor et al. 2020: 31707).

In Canada, this concentration is even more pronounced as only three slaughter and *meatpacking* plants are responsible for processing 95 per cent of all *beef* (Mosby & Rotz 2020). This points to a serious issue in the Canadian food supply chain because, as Mosby and Rotz argue, "our food system has been transformed to disproportionately benefit massive multination corporations like Tyson, Cargill and JBS at the expense of farmers, workers and – as we're now seeing in the form of empty grocery store shelves and steadily rising food prices – consumers" (Mosby & Rotz 2020).

There are additional racial, class, and gender dimensions to working in a slaughterhouse. *Poultry* slaughter and *processing* workers in the United States are often vulnerable because of their status as undocumented workers and/or a lack of English-language proficiency (Arcury et al. 2015). In a study of mainly African American women employed in a variety of low-wage jobs in North Carolina, those in chicken *processing* reported significantly higher depressive symptoms than those employed in other areas (Lipscomb et al. 2007: 293).

In an ethnography of one of the largest chicken slaughterhouses in the world, Steve Striffler found that most of the workers on the processing line were women, even when two-thirds of the overall workers were men. The line work is considered to be the worst in the factory as it is monotonous and fast paced. Workers responsible for hanging chickens on the conveyer line processed around 40 chickens a minute (2005: 115). This type of repetitive motion is associated with the development of repetitive strain injury (Striffler 2005: 115; Arcury et al. 2015).

Undocumented workers face even more severe health outcomes from such injuries because they may also lack access to health care (Arcury et al. 2015). Research on *pork* slaughter and *processing* found employees who work with live animals are more likely to experience wounds, such as cuts and scratches, and this increases their chance of infections (Kyeremateng-Amoah et al. 2014: 678–9).

The way to reduce the risk of injury for workers is to slow down the line speed, limiting the repetitive motions a worker must make, and ensuring that the tools used are sharp (Arcury et al. 2015). Workers, like farmers and those working in slaughterhouses, who come in contact with live nonhuman animals and/or their bodies or excrements, may be at a higher risk of exposure to pathogens (You et al. 2016: 462), especially those in the *pork* and *poultry* industries (Porcher 2011: 9). Based on the risks of slaughterhouse work, Jennifer

Dillard argues that it might be suitable to legally frame it as an "ultra-hazardous activity," and the workers be eligible for compensation (Dillard 2008).

The harms present in intensive *farming*, slaughter and *processing* facilities extend beyond the workplace to the communities in which they are located. As a result of pig production in the United States, particularly in the southern states for instance, airborne emissions, water pollution, and human health risks exist in rural communities, and have serious implications for "African-American residents and farmers ... [who] have been bearing more than their fair share of the hog industry's pollution, economic costs, and land displacement pressures" (Ladd & Edward 2002: 29).

Compartmentalization of Human Work Facilitates De-animalization

De-animalization refers to the process whereby nonhuman animals are treated as though they are products or commodities (Hamilton & McCabe 2016; Harfield et al. 2016). The organization of slaughter facilities, as well as the way that such *food* is packaged and marketed to consumers, helps to facilitate this transformation. The organization of slaughterhouses is part of a process that sanitizes the killing of nonhuman animals because the compartmentalization of tasks, and the architecture of the facility, separates most workers from the actual killing of animals (Hamilton & McCabe 2016; Pachirat 2011).

Some work in slaughterhouses may be explicitly labeled, such as "lower belly ripper" and "internal fat cutter." However, other labels such as "knocker" (the worker who uses a captive-steel bolt to stun the cows) and "rim over" (the person responsible for removing the cow's skin from their shoulders) hide the violent realities of the particular job (Pachirat 2011: 257–70). But throughout the highly automated process, the nonhuman animals start to look more like *food* than their former living selves (Hamilton & McCabe 2016: 340).

The "politics of sight" asks us to question certain features of the slaughterhouse industry, such as how is it organized, who is likely to work there, and how do the distances between this work and consumers hide the realities of what is happening? As a concept, it relates to the efforts to try and make visible those hidden aspects of slaughterhouses to facilitate positive change (Pachirat 2011: 14–15).

What we see and what we choose to ignore relates to social and political power.

In thinking through how slaughtering can be "hidden in plain sight," we are asked to think about some of the barriers, or distance, such as those related to precarious employment: race, gender, immigration status, and educational levels (Pachirat 2011). Consumers essentially know that *meat* comes from other animals, but they can hide behind the knowledge (and distance) that someone else is performing the work that they themselves may find dirty, demeaning, and dangerous.

It is not only in these *meat* and *chicken* slaughter and *processing* facilities that compartmentalization and de-animalization occur. Just think about a trip to your local supermarket. *Food* products made with nonhuman animal ingredients often look nothing like their former selves, and people most often buy *meat, fish,* and products made from nonhuman animals in wrapped packages covered in plastic, and labeled with names that may also obscure what is being consumed, such as *hamburger, veal, pate, sushi, cheese,* or *omelet* (Adams 1991; Stibbe 2001; Nibert 2002; Smith-Harris 2004; Harris 2017).

Carol Adams argues that "through butchering, animals become absent referents. Animals in name and body are made absent *as animals* for meat to exist ... Without animals there would be no meat eating, yet they are absent from the act of eating meat because they have been transformed into food" (1991: 40). As animals in the *food* industry, it is not only the slaughtering process that commodifies them, but they are also commodities even during their short lives. Consequently, they embody a strange in-between category that while they are alive, the end product of their labor (becoming a *hamburger* or *fish stick,* producing *milk* or *eggs*) justifies all sorts of conditions and treatments that are not seen as acceptable (or legal) for other categories of nonhuman animals that we hold dear to our hearts, such as companion animals.

Garry Leech, who worked as a butcher in a supermarket for eight years, explains how he did not view the slabs of *meat* and *carcasses* that he dissected as nonhuman animals:

> When I think back to that work I realize that I never viewed the meat I was cutting up as a former living creature; a sentient being. It was just meat. I had grown up in a society disconnected from nature, including food animals, and so I only saw them as product, as food. My urban upbringing had resulted in me becoming alienated from most nonhuman

animals, and from nature in general. But it's important to point out that I enjoyed my job, as did my fellow butchers. Probably in the same way many coalminers can't imagine doing any other kind of work, even though most of us can't imagine engaging in such dangerous labour. In other words, not everyone who engages in this sort of gruesome or dangerous work feels exploited. (Personal correspondence with Garry Leech 2023)

Leech's comments show how the de-animalization process is already occurring before the *meat* is packaged and put on display for consumers to purchase. The packaging of the *product* just takes this process one step further.

Dirty Work and Slow Violence

During the COVID-19 pandemic, *meat processing* workers in both the United States and Canada were labeled as "essential workers." While food is essential for life, it is difficult to demonstrate how the production of *meat* could be essential to the national security of the United States, and as such brought under The Defense Production Act. As we discussed earlier, the Select Subcommittee on the Coronavirus Crisis maintained that even the companies and governmental agencies that made this case knew that closing meat processing plants was not a threat to food security, only to corporate profits (US House of Representatives 2022).

In Canada, work in slaughter and *packing* plants, along with other food production, transportation, and food sales, were defined as an "essential service," and as a result of this fact "workers who deliver essential services and functions should continue to do their jobs provided they have no symptoms of COVID-19 disease" (Government of Canada 2021b). Pursuant to this stance, the federal government in Canada issued funds for slaughterhouse companies under the Federal Emergencies Act, in part to pay for personal protective equipment and factory modifications such as plastic barriers.

COVID-19 added another threat to work that is already dangerous, has a high turnover rate, and often low wages (Broadway 2013; Haedicke 2013: 123–4). High turnover creates additional safety issues as it results in a perpetually inexperienced workforce in an already dangerous work site (Fitzgerald 2010: 64). Haedicke argues that "[t]he fact that meatpacking workers currently have poor wages, high injury rates, and very unpleasant working conditions is a product of conditions that erode workers' ability to collectively challenge corporate market strategies that generate profits by keeping labour expenses low" (Haedicke 2013: 128).

In their study on *chicken* slaughter and *processing* workers, Mora et al. conclude that "workers' willingness to endure dirty, dangerous, and demanding (3-D) conditions" must be examined within the context of what aspects of the job were seen as beneficial to some of the participants in their study, such as decent pay (relative to other jobs the workers could find), benefits (such as health benefits and overtime), and reliability (the company always paid their wages on time and there were limited other positions they could access in the community) (2016: 877–81). Workers identified the physical aspects of work as problematic, such as the fact that it was exhausting, dirty, and physically demanding. They also pointed out their health concerns, such as pain from the work, and negative relationships with supervisors and coworkers, such as a lack of autonomy and feeling pressured to maintain the line speed (Mora et al. 2016: 881–2).

The seriousness of these vocational challenges is generally exacerbated by the fact that the workers in this study were primarily Latino immigrants who had few other options for stable employment and could not easily quit their jobs even when faced with the "dangerous, dirty, and demanding" conditions of employment (Mora et al. 2016: 882–4). Although some workers identified certain aspects of their jobs that they appreciated, such as decent pay via-à-vis other work available in the community, an overall milieu of structural violence was identified by the researchers as a central reason that these workers stayed in these jobs.

As McLoughlin notes, from a worker perspective, some jobs within a slaughterhouse are better (and worse) than others (2019: 328). Some workers in their study of an Irish slaughterhouse identified that work in the "lairage" (working directly with the animals, such as herding and putting them in pens) was preferable, as opposed to working on the line, because it involved more freedom of movement and the ability to talk with co-workers (McLoughlin 2019: 328).

"That shit will fuck you up for real"

In the slaughterhouse industries in particular, workers face immediate physical dangers due to the speed with which the work proceeds (Dillard 2008; Haedicke 2013). Perhaps less well known are the negative implications for workers' mental health, and these outcomes are more pronounced than for workers in other industrial settings (Baran et al. 2016). This is thought to occur because of the routine violence workers must engage in, and the importance of further consideration of "perpetration-induced traumatic stress" (PITS) (Baran et al. 2016: 364). Workers may experience PITS because of their role in killing nonhuman animals. PITS results not from an individual experiencing a severe trauma such as their life being in imminent danger, as is often the case with post-traumatic stress disorder (PTSD), but from their role in causing harm to another living being (Whiting & Marion 2011).

In Pachirat's ethnography of a slaughterhouse he describes trying the job of "knocker," where he used a captive bolt to stun several cows. He depicts the stunning process this way:

> I am so focused on the gun that I do not even notice the animal that comes through on the conveyor. Its head swings back and forth widely, eyes bulging. Then it stops moving for a moment, and I hold the gun against its skull and pull the trigger ... The gun recoils in my hands, and I see a hole in the animal's skull. Blood sputters, squirts, and then begins flowing steadily from the hole and the animal's eyes roll up into its shaking head. Its neck is extended and convulsing, and its tongue hangs out the side of its mouth. I look at Camilo, who motions for me to fire again. I shoot, and the animal's head falls heavily onto the conveyor below. (2011: 150)

The description Pachirat offers seems unemotional, but it would be a mistake to infer that the job on the kill floor was just another job in this slaughterhouse. Pachirat goes on to discuss the reaction of fellow workers to his request to work on the kill floor. It is in their reaction that we learn more about this work, the perceptions of other workers towards the workers who kill nonhuman animals, and the feelings of workers towards the act of killing itself. One co-worker asks, "'Why you out there doing that? You want to be the knocker?' When I say maybe, he responds, 'No, you don't want to do that. I don't want to do that. Nobody wants to do that. You'll have bad dreams'" (Pachirat 2011: 150–1).

Another worker explains, "I already feel guilty enough as it is ... Especially when I go out there and see their cute little faces" (2011: 151). A third worker implores Pachirat to reconsider working on the kill floor and tells him, "Man, that will mess you up. Knockers have to see a psychologist or a psychiatrist or whatever they're called every three months ... Because man, that's killing," he said; "that shit will fuck you up for real" (Pachirat 2011: 152–3). It is clear from these reports and anecdotes that workers experience trauma from being active participants in causing harm to living beings through the act of killing nonhuman animals (Dillard 2008).

We had an opportunity to interview former slaughterhouse worker and retired teacher Paul Rooke about his experience several decades ago in a small-scale, yet still mechanized, Australian slaughter facility. One of the things he told us was that this experience remains with him all these years later. Rooke discussed this trauma and explained, "I find it even difficult to talk about now." He also confessed to being "a little bit edgy" leading up to the interview (Interview with Paul Rooke 2023).

Hamilton and McCabe discuss the complexity of emotions that *meat* inspection workers demonstrate but maintain that the way in which the work is organized and the slaughter facility itself is laid out eliminated any expression of emotions in the facility (2016: 345). In our interview, Rooke discussed his co-workers, who were all men, as having two different personalities, the personality at work, and the personality outside of work:

> While they were working and how they were talking and how they were behaving as the sort of hardened individuals. And then when they got out, went to the pub and they became human again and it was like dealing with two different kinds of personalities, two different kinds of people. And I loved their pub personality where they could talk about what they loved and were interested in and whether it be sport, music, whatever, we would have these fantastic, intense conversations outside of work, but inside it was a different world. It was like they had to be a different kind of person and they knew that they did that until the end of the day. (Interview with Paul Rooke 2023)

Rooke also mentioned that alcohol factored prominently in the day-to-day life of workers, saying that it was routinely consumed before work, at lunch, and after work. This claim is consistent with a study of Danish workers across a variety of occupation types, which found that slaughterhouse workers had higher alcohol intake on weekdays and weekends when compared to employees in other occupations (Baran et al. 2016: 362).

When we asked Rooke to describe his experience working at the slaughterhouse, he explained that his time there was "short but profound":

> So, the heat, the smell, the blood, the cries of the animals, the ongoing horrific nature of the process, the building, I mean, it haunts me … It was only three months of my life as opposed to years, and it was profound. I still describe it as profound and for years I couldn't stand the smell of lamb in particular and if I smelled it, and knew about it, it would repulse me, and I didn't of course, I shouldn't say of course, but I didn't eat meat for years … But even then, it was red meat in particular … it just as I say had a profound effect on me. I often described it in terms of the senses. So, when you're in a place like that, what you see and what you smell and what you hear, almost what you taste in your mouth, it just stays with you. (Interview with Paul Rooke 2023)

Research suggests that normative ideals of masculinity are emphasized within slaughterhouse work, and that this process of suppressing feelings and emotions likely involves "negotiation and repression of emotion" for workers (McLoughlin 2019: 332). Some have portrayed slaughterhouse employment as "dirty work" but emphasize that it may differ from other types of employment constructed in this way precisely because it involves "a different set of moral norms" and it requires doing things to sentient beings that would not be allowed in a human-to-human context (Tallberg & Jordan 2021: 860).

When comparing slaughterhouse killing to other types of killing nonhuman animals, such as hunting, it is found to have three unique features: it is routinized, it is experienced close up, and it involves a different set of relationships because the nonhuman animals are domesticated and have a relationship of trust with humans. For these reasons it may be more troubling for workers (Baran et al. 2016: 355–6). When we asked Rooke to tell us a bit about his experience working there, the description he gave was not of a typical day, but it was a memory that stayed with him all these years later. He explained:

> As I say, the experience with pigs will never leave me. So having said that, there were certain instances that will stay with me. For example, one day a pig fell from the initial hook after its throat was cut, into the pit, which meant somebody would have to go down and get that animal … The men at that point sort of revealed how hardened they were to the process and they were chastising the foreman because he was unwilling to go down and get this pig hooked back up. And so, another man had to go down with a metal club and kill the animal as we watched so that it could be rehung and continued on in the process [as] of course time is money, and they need to get that process going again. (Interview with Paul Rooke 2023)

"Dirty work" involves work that is seen as undesirable and that can be demonstrated in the above response of the manager, which showed his hesitancy to be the one to go and kill the pig. But dirty work is more than simply undesirable; it is also tainted or stigmatized because of the physical matter involved (i.e., blood, intestines, fecal matter) and the "taint" of those who cause pain, suffering, or death to another (Baran et al. 2012; 2016: 351–4; Tallberg & Jordan 2021: 860).

While we can imagine how this stigmatization can be problematic from a mental health perspective for human laborers who work in such industries, this is also part of the disconnect between consumers of products made from nonhuman animals and the process of turning live nonhuman animals into the final *food* product. Someone else literally does the dirty work so that consumers can peacefully enjoy eating *meat*. Additionally, the way that the factory is organized, the corporate concentration of the slaughterhouse industry, which involves centralized factories often in small, remote communities, ensures that most communities and consumers are shielded from the realities of this process. At the end of the day, "animal agriculture is foundationally a violent institution that entails routine and spectacular intra-human, inter-species, and environmental harms" (Struthers Montford & Wotherspoon 2021: 84).

The conditions for workers in a slaughterhouse often include trauma-inducing work and a "slow violence" that extends beyond the walls of the slaughterhouse, because the workplace sometimes creates environmental conditions that contaminate neighboring communities with the "deadstock" of the factory *farm* output or from the disassembling process (see Struthers Montford & Wotherspoon 2021). Slow violence, a term coined by Rob Nixon (2011) in his book *Slow Violence and the Environmentalism of the Poor*, and used to describe slaughterhouse work during COVID-19 by Struthers Montford and Wotherspoon (2021), is defined as something that "occurs gradually and out of sight, a violence of delayed destruction that is dispersed across time and space, an attritional violence that is typically not viewed as violence at all" (Nixon 2011: 2).

The phrase clearly relates to the expendable nature of many workers within the slaughterhouses, and their higher-than-normal rates of COVID-19 infection within this industrial setting. It is also a useful concept in helping us explore the environmental issues created by factory *farms* and slaughterhouses that were discussed earlier in the book. And as we will see below, slow violence helps to explain the seeping of other interpersonal and community-based issues into the locales and communities where they are located.

The Spillover Effect

As we discussed above, killing of nonhuman animals in the slaughterhouse industry has been found to have negative emotional consequences for human workers. It also has consequences for the larger communities where such operations are situated. In chapter 2, we discussed the negative externalities, or spillovers, related to the environmental toll large-scale industrialized *farms* often have on surrounding areas and waterways.

In this chapter, "spillover" is related to the human costs of slaughterhouse work, such as increased arrest rates for "violent crimes, rape, and other sex offences" in US counties where such workplaces exist (Fitzgerald 2015: 108; for a detailed overview of the study, see Fitzgerald et al. 2009). When we interviewed sociologist and green criminologist Amy Fitzgerald, we asked how she would explain the concept of "spillover" to someone unfamiliar with it and she responded:

> Well, to my way of thinking, it is looking at something, and it could be many different types of things that you expect to be contained to one area, and then they transcend these arbitrary boundaries and then it is part of another area that you would not have necessarily expected. So, there is this spillover from something to another area that you would have expected to be divergent, but there's actually a crossover there. (Interview with Amy Fitzgerald 2022)

In this case, the spillover she and her colleagues found relates to the level of arrests in communities with slaughterhouse employment (Fitzgerald 2015: 108). They found that "particularly with violent crimes ... there was a spillover effect from employment levels in slaughterhouses in the county to crimes against people. So, the spillover from harming animals to then harming people. As best as we could tell ... after we controlled for so many different variables that are correlated with crime, we still found that impact of the spillover of the slaughterhouse employment" (Interview with Amy Fitzgerald 2022).

In a more recent study also utilizing Uniform Crime Reports (UCR), sociologist Jessica Jacques studied nonmetropolitan communities in the US with "the highest concentration of beef slaughterhouse facilities and employment" in such occupations (Jacques 2015: 599). She found that such communities had higher total arrest rates, and rates of arrests for particular crimes such as rape and those involving the family (Jacques 2015).

It is important to note that such studies have limitations as they are in aggregate form so that it is impossible to know precisely which individuals are committing the crimes (Fitzgerald et al. 2009: 175; Fitzgerald 2015: 108). Also, relying on UCR presents additional limits as it relies on reports to police, and some crimes are known to be underreported (such as sexual assault/rape) (Fitzgerald et al. 2009: 175). Furthermore, there may be variations in reporting because of differences in police resources and practices across regions (Jacques 2015: 609). However, there are still important opportunities for further research here, as we must employ a precautionary principle in considering the ways in which having a slaughterhouse in a community may be connected to increased crimes against humans in that locale (Fitzgerald et al. 2009; Jacques 2015).

Deskilling and the "Entanglements of Oppression"

Inevitably, there exist relationships between human workers and the nonhuman animals they encounter in the industrialized *food* production system. Porcher argues that "there is no such thing as 'industrial animal husbandry'. This expression is an oxymoron that suggests that even animal farming could survive and still make sense in an industrial system. However, studies show that the dual injunction of rearing and producing is impossible to reconcile and is a major cause of suffering for both workers and animals" (Porcher 2011: 14).

This is because the main impetus of an industrial system is production and profit, which turns nonhuman animals into a commodity, thereby often ignoring both the well-being of human workers and other animals in the process. In considering the experience of human and nonhumans within this industrial *food* system, we turn to Barbara Noske's (1997) re-working of Karl Marx's concept of alienation and David Nibert's (2002) cross-species understanding of "systematic and entangled oppression."

Marx's concept of alienation can be useful in thinking through the realities for slaughterhouse workers, and other workers at the various stages of industrialized production of nonhuman animals as *food*. Marx conceived of a structural condition under capitalism, in which workers did not have control over their work and in which they inevitably ended up feeling powerless and disconnected from the product (in this case, nonhuman animals under intensive production), the productive activity (killing and *processing* nonhuman animals in the slaughterhouse), their co-workers, and even themselves. At the end of the day, such alienation leads to feelings of misery rather than well-being (Noske 1997).

Noske discussed the deskilling that happens in the A-IC from the nonhuman animals' standpoint. Borrowing from Marx's concept of alienation, she creates the concept of "de-animalization." Under the factory farming system, nonhuman animals are alienated from their own bodies, their productive output, others of their species, nature, and species life in general (1997: 18–20). Both Marx's concept of alienation and Noske's de-animalization can be useful in helping us to understand the A-IC, especially in terms of how workers may come to act as machines, and nonhuman animals are constructed as commodities. Karen Davis applied Marx's concept of alienation to the lives of chickens under intensive agriculture. Davis argues that chickens are alienated from their own products, for instance, which includes their eggs, offspring, and even their own bodies as they are degraded through genetic modifications to ensure they produce as much muscle (*meat*) as possible, as quickly as possible (2014: 175).

In order to achieve changes in the speed of raising nonhuman animals as *food* and keeping prices low, major shifts in how these animals are housed were instituted. These changes mean that many nonhuman animals used in the industrial production of *food* now live their entire lives indoors, experience increased crowding (such as *battery cages* for egg-laying chickens and *gestation* and *farrowing crates* for pregnant and nursing pigs) or in isolation (such as calves held in individual pens in some *veal* production). Nonhuman animals often suffer painful modifications to their bodies which are deemed necessary to enable them to live with others of their kind in such close and unnatural confines. Chickens, for instance, are routinely *debeaked*. This involves searing off part of their beaks so they cannot peck each other when housed in a tight confined space (Davis 2014: 178).

Pigs routinely have their tails cut off, euphemistically referred to as *docking*. Again, this is necessary only because they are being placed in artificially cramped quarters to which they may react by biting each other's tails. Isolation also occurs in industrial *hog* production through the use of *gestation crates* for pregnant and nursing mothers (for a detailed description and photos, see Blanchette 2020: 74–8; 122–6). In these, mother pigs are confined in a space about the size of a refrigerator and can stand and lie down but cannot turn around. The industry claims that this is for the safety of the piglets, so that they do not get crushed by their mother. But if they were given more room than is allowed within the factory system, this would not be a significant concern (Medora 2014).

Isolation is also a significant factor in other animal industries, such as the *dairy* industry. Within this system the product is obviously the mother's milk. In the *dairy* industry, newborn calves are typically taken from their mothers within a few hours (typically from 30 minutes to 12 hours) of birth and fed a milk supplement (Vasseur et al. 2010; Pempek at al. 2017), so that their mother's milk can be sold to human consumers. Free-living mammals with hooves naturally wean gradually over a period of several months to a year (von Keyserlingk & Weary 2007: 110). When cows have control over their own bodies and lead self-determined lives, females remain with their mothers for the duration of their lives, and males do not break away from the herd until they are around one year old (Berreville 2014: 188). Cows in the *dairy* industry are not permitted to nurture their calves or forge the significant mother–baby bond.

Some may ask, but what about eating fishes? That must be a better option, right? Unfortunately, there exists a series of profound issues with eating fishes as well. The documentary *Seaspiracy* outlines some of the issues with intensive fishing, such as injury and deaths of enslaved workers, depletion of wild fishes, and the death of other marine mammals as *by-catch* (2021). The United Nations Food and Agriculture Organization describes the commercial *fisheries* as one of the most dangerous occupations in the world (Food and Agriculture Organization of the United Nations 2023).

Jonathan Balcombe, Director of Animal Sentience at the Humane Society Institute for Science and Policy, writes that *fish-farms* are a "subset of aquaculture" and it "is the fastest-growing animal-*food*-producing sector in the world, having gone from 5 percent of global fish production in 1970 to about a two-fifths of the total today" (2016: 214). These so-called *fish farms* hold the fishes in "marine and freshwater net pens, or land-based tanks or ponds" (Balcombe 2016: 214).

While some may see intensive *fish farming* as a way to avoid some of the issues outlined above, confined fishes experience "a mundane, unstimulated life of captivity [that] stunts brain development and function. When hatchery-reared fishes are recaptured after release into the wild, their stomachs are often empty or filled with inanimate objects such as floating debris or stones that look like pellets they were reared on" (Balcombe 2016: 216–17).

The density of so-called *farmed fish* also creates several serious issues, from an individual animal's health perspective, for consumer health, and for other species and their ecosystems. Crowding means that if the fishes have an infestation, such as sea lice infections, they cannot move away and are likely to infect another nearby fish (Hume 2004; Balcombe 2016). This parasite eats "through the mucus, flesh, and eyes of fishes" in confined conditions. Within confined fish operations, "death rates of 10 to 30 percent are considered acceptable" (Balcombe 2016: 216). It is important to note that the spillover from these confined fish operations leads to environmental and species-specific die-offs for wild fishes that live in the vicinity of these concentrated feeding operations for fishes (Hume 2004). As Jonathan Balcombe explains:

> The nets that hold fishes in their sea pens do not prevent these rampant parasites from escaping. A female sea louse lays about 22,000 eggs during her seven-month life span, and they spread in clouds over miles of surrounding waters, wreaking havoc on wild fishes who find themselves in the vicinity of the farms. The lice are credited with causing massive die-offs of 80 percent of wild pink salmons on Canada's Pacific coast. Trickle-down effects impact salmon-dependent wildlife: bears, eagles, and orcas. (2016: 216)

In West coast salmon *farming*, it is believed that over 1.4 million *farmed* salmon had escaped into the Pacific Ocean (at least 500,000 of which were Atlantic salmon being held in Pacific waters) prior to 2004 (Hume 2004: 50–1). Some of these fishes were found as far away as the Bering Sea and "[c]ritics fear that farmed Atlantics could open a disease vector into wild populations with no natural immunity and that drugs used to keep farm fish healthy might spawn antibiotic-resistant supergerms in the ocean" (Hume 2004: 51).

The alienation of sentient beings from the lives they could be living becomes more obvious when we see how they behave when outside the industrial system. Nonhuman animals, such as wild-living fishes, or other animals rescued from factory *farms* and placed in animal sanctuaries, often exhibit behaviors that point to the richness lost in the intensive production of other animals. In these situations, they can form complex relationships with others, take part in meaningful behaviors such as foraging for food, spending time outdoors, playing, and so on.

The treatment of nonhuman animals raised and killed in the A-IC, as well as the workers charged with killing and *processing* them, points to intricate intersections of oppression. In taking a sociological approach to the oppression of both human and nonhuman animals, David Nibert put forward a theory of oppression to frame their "systematic and entangled oppression" (2002: 14). He argues that in order to perpetuate oppression, there are three factors needed: economic exploitation, unequal power, and ideological control (Nibert 2002: 15).

Economic exploitation can be seen in the A-IC generally, and in the slaughterhouse industry specifically. As Nibert tells us, "humans tend to disperse, eliminate, or exploit a group they perceive to be unlike themselves" (Nibert 2002: 13). The workers employed to kill and *process* nonhuman animals are often exploited under this system as they are paid low wages and expected to work under stressful and dangerous working conditions. Unequal power, usually connected to the ability to "exercise political control [and] wield the power of the state," can be seen in the case of corporate concentration of slaughterhouse industries (Nibert 2002: 13). Finally, ideological control facilitates prejudice and discrimination, which makes such social, political, and economic arrangements "appear natural, thus making the oppression invisible to those who enjoy privilege and who gain some benefit from such oppression" (Nibert 2002: 13).

The labor of both slaughterhouse and farm workers in this intensive system remains mostly invisible to those on the outside. And the conditions of nonhuman animals under this system is rarely known outside of the industry, and when consumers do hear about it, it is presented as a necessary and inevitable part of food production. This is possible because socialization normalizes the process of production as well as *meat*-heavy diets. As Nibert tells us, "[t]he oppressive treatment of immigrants and other animals, which is entangled and inextricably tied to the capitalist system, directly intersects in the nation's slaughterhouses" (2014: 3–4).

In the commodification of animals and their output, the comfort and well-being of individual nonhuman animals will necessarily come second to potential profits. This is the logic of capital. It is important to think through the relations that allow for and facilitate this system, as well as the implementation of efficiencies that benefit the bottom line but often fail to consider whether they also reproduce existing societal inequalities and inequities.

Recognizing Animals' Labor

In researching this chapter, we found that the issues facing human workers, such as increasing speeds of production lines or the expectation for them to suspend compassion, also have implications for nonhuman animals during *production* throughout the A-IC. As we saw previously, the shift in the mid-twentieth century to industrial *animal agriculture* intensified the methods for taking live animals and turning them into *food*. Under this system virtually no area of these animals' lives, and we could say labor, goes untouched by humans.

They are artificially inseminated, fed diets that they would never consume if left to their own devices, and their life cycles are sped up so that they grow larger faster, and are sent to slaughter in a much-reduced timeframe (Fitzgerald 2015). This means that we now have "a meat-production system that can turn a fertile egg and a nine-pound (four-kilogram) bag of feed into a five-pound (two-kilogram) chicken in five weeks" (Patel & Moore 2017: 156). Such nonhuman animals often live these short lives inside, unable to perform many of the activities that make for an enriched and, we would add, joyful life, including caring for their young or socializing with others of their species (Noske 1997; Berreville 2014). Instead of foraging outdoors, *sows* are now "bred for quite radical forms of docility, such that they can survive in tiny [gestation] crates without exhibiting behavior that is *excessively* debilitating" (our emphasis) (Blanchette 2020: 138).

The founder of United Poultry Concerns, Karen Davis, points out how domesticated chickens that would have had a life expectancy of up to 10 years can now expect to live only days, weeks, or months because they are bred to produce disproportionately large breast tissue that hampers the development of their organs, muscle, and the functioning of their limbs. Even chicks rescued when days-old often develop genetically influenced diseases, such as those related to the cardiovascular and gastrointestinal systems, as a result of their genetic modifications (Davis 2014: 173).

Years ago, as a summer student employee, Tracey worked as a soil researcher at an agricultural college. One of the central tasks of this job was traveling to local farms and conducting soil density tests to help measure a farm's soil erosion and to strategize on methods to counter this risk. Some of the farms visited were small, non-intensive *dairy* farms. She tells the story of encountering a cow in the field:

> As we busily took soil samples and measured how quickly the soil could absorb water, the cows often grazed alongside us. One day, as I was sitting in the field taking samples, a cow came up behind me and licked the back of my neck. I remember being startled by this action and later when I was talking to the farmer and retold the encounter, he knew exactly who I was talking about and gave me other examples of this cow's friendliness and her affectionate personality. At that time, as a young student research assistant, I never thought about this cow as being a labourer, or the way in which her labour was necessary for me to have milk in my tea, or how when she was no longer producing milk, she would be sold at a livestock auction as a cull cow, shipped to a slaughter facility, and her body would be used to make hamburgers or processed foods. (Harris, personal notes)

Thinking about this cow now, from a labor perspective, opens the door to many interconnected and important concepts that we have begun to unpack in this book: transparency, empathy, de-animalization, the gendered nature of much animal labor (which will be discussed in chapter 5), the intersections of oppression, as well as the structural violence that workers, both human and nonhuman, face under the modern capitalist production of food. Thinking about nonhuman animals in the *food* system as laborers may be important for shifting perceptions of them and their inherent value:

> [M]ost animals exist today in food systems and experimental labs solely to become use values for humans (profit and food). This is one of the reasons I think it useful to understand animals today as workers under capitalism, as it helps to frame the problem in a different way. Situating animals as labourers forces us to recognize animals as agents of production; agents who have an interest in avoiding the painful, exploitative and deadly work they have been compelled to do, work that by its nature diminishes their prospect of leading flourishing lives. (Blattner et al. 2021: 263)

When we interviewed Brandon Keim, an independent journalist specializing in nonhuman animals, nature, and science, we asked about his understanding of other animals' labor. Brandon framed his answer by first giving credit to Brock University professor Kendra Coulter's work (see *Animals, Work, and the Promise of Interspecies Solidarity*, 2016). He then explained that, "clearly animals work for us, that's so obvious and unavoidable, and I think one of the powerful things about that idea is it asks you to notice that. Because the animals work for us, but we don't notice it, and we don't think of it as labor until maybe somebody puts it in those terms" (Interview with Brandon Keim 2022).

Once we are asked to recognize this labor, and it is brought to our attention, it reframes the way we can consider nonhuman animals in the *food* industry. Considering their labor may create the possibility, as Charlotte Blattner states, for us to ask some difficult questions: "What exactly do animals want? Are just short or long-term relationships between us possible? What types of human–animal relationships are acceptable to them? How can we make sure that we are not using the animals' interests as an excuse to advance our own, but that instead we are truly working toward relationships that empower us both?" (2020: 50).

This acknowledgement of nonhuman animals' labor would have to be carefully positioned so that it actually considered what is in the best interests of other animals, and not used as a way to justify their continued exploitation (Canavan 2017). As political scientist Jana Canavan cautions, if nonhuman animals were actually making a conscious choice to participate in farming processes, the A-IC would not have to create confinement methods (e.g., *gestation crates*) or practices that allow nonhuman animals to be used in particular ways (e.g., removing babies from their mothers quickly after birth in the *dairy* industry) (2017: 44–5).

Kendra Coulter argues that we need to create "humane jobs" as an alternative to dangerous and unsafe work conditions, while recognizing that these jobs currently exist within a system that also harms nonhuman animals and causes environmental crises (Coulter 2020). Coulter considers the need for humane job creation, the facilitation of improved conditions of work in some already existing areas, and whether there is the possibility of nonhuman animals participating in some of those humane jobs (Coulter 2020: 29–30). She is clear in her analysis of what this might look like: the humane job must not harm animals, and so the commodification of animals would not constitute a good job or *humane* job (Coulter 2020: 35). We asked Brandon Keim what compensation might look like for nonhuman animals' labor. He replied:

> My rule of thumb answer is that it would look like a genuinely good life. So, I know people who raise animals for the food industry or for food will say that they're feeding the animals, they're providing medical care, they're protecting them from predation, doing all these things. I think that's absolutely true, and I think that is compensation, but I think as it is right now, it is just not nearly enough. The lives of those animals are still extremely short, there's all kinds of other things that are important to them that are not provided. And so I think, fair compensation would look like a genuinely good long life with an intact social structure with rich social relationships, and all of the species-specific things that make an animal's life one that, if one of our loved ones was to be reincarnated as a chicken or a pig, you would be okay with that. I think, to me, that's sort of the litmus test of it. So yeah, I think that's what compensation would look like. And there's some animals who do other forms of work like transportation or lifting or therapy and so on, where compensation for them would also include a retirement and in that period of time at the end of their lives they're just living for themselves. (Interview with Brandon Keim 2022)

Considering the labor, and indeed complex relationships and lives of nonhuman animals, from this kind of holistic vantage point would require dramatic shifts in how they are treated. When we asked Margaret Robinson, Mi'kmaw scholar, member of Lennox Island First Nation, and Canada Research Chair in Reconciliation, Gender, and Identity, about the possibility of giving voice to other animals, she said:

> I am hesitant to see humans as having the role of being able to voice the needs and preferences of animals because I think that could very easily be misused. However, I don't think you need much imagination to be able to figure out that animals also need clean drinking water. Animals also need to have access to their traditional food sources. The things that I would affirm for a neighbor should be applied to my animal neighbors and I think we can argue on behalf of their rights as persons without necessarily having to understand how they themselves would express those rights. (Interview with Margaret Robinson 2022)

We can see that the kinds of needs articulated in this argument are counter to profit maximization and that they are the types of things that a standardized industrial food production system is unable to provide. In order to consider such needs, we would be required to reimagine nonhuman animals as individuals with their own needs, desires, frustrations, and joys – or in Robinson's words and research, as "persons" (Robinson 2014). In doing so, the possibility of continuing to utilize them as commodities and living property would become much more difficult to contemplate.

Transparency and Beyond: Cultivating Empathy?

At the beginning of this chapter, we had suggested that many readers might struggle to imagine what it looks like to work in the industrial production of other animals for *food*. What does work in a slaughterhouse look, feel, sound, and smell like? What about in a chicken hatchery? Or driving a transport truck loaded with live animals on route to a slaughter facility? What does work in the commercial *fishery* or with fishes in an *aquaculture* facility entail? And fewer people still may have stopped to consider the labor of nonhuman animals in this industrialized system.

In order to sketch out some answers to these important questions, we began to shed some light on these industries and the labor that is involved in the production of nonhuman animals for *food*. This pulling back the curtain, so to speak, is really offering up transparency. Transparency is important because knowing what is happening in these spaces and places will allow us to make conscientious choices about what we eat (Lund et al. 2007; Singer & Mason 2007; Fitzgerald 2015). If we pause to think, we may be concerned about why we know so little about the labor required to turn nonhuman animals into *food*, we might wonder about the workers whose labor is so essential (both human and nonhuman), and we might want to know what these spaces look like or where they are located.

When we spoke to sociologist Michael Haedicke about the importance of a transparent food system, he explained that:

> When we talk about transparency, we tend to think about it in terms of knowledge, of understanding what happens and being able to see and observe what happens at different points in the food chain. But I think there also has to be, along with transparency, ways to cultivate empathy or ways to cultivate not only knowledge but also a deeper understanding of people's experiences, which can be difficult to do because of those barriers and those gaps ... So to say I understand what goes on in this

type of farm is different from saying I have some insight into the experiences of people in this type of farm, right? The frustrations that [they experience], the benefits, the joys, and the sorrows. And I think that sort of understanding is one that is more essential to cultivate beyond simple rational knowledge of what happens at other points in food systems. (Interview with Michael Haedicke 2022)

Professor of Philosophy and Coordinator of Wesleyan Animal Studies at Wesleyan University, Lori Gruen advocates for an "entangled empathy" (2015: 2). In proposing this ethics of care, she explains that entangled empathy is "a type of caring perception focused on attending to another's experience of wellbeing. An experiential process involving a blend of emotion and cognition in which we recognize we are in relationships with others and are called upon to be responsive and responsible in these relationships by attending to another's needs, interests, desires, vulnerabilities, hopes and sensitivities" (Gruen 2015: 3).

Josephine Donovan, Professor Emerita at the University of Maine, and independent scholar Carol Adams explain that this "ethic-of-care" differs from, say, an "animal welfare approach," because in addition to focusing on the suffering of individual nonhuman animals, attention is also paid "to the political and economic systems that are causing the suffering" (2007: 3).

Perhaps cultivating this type of empathy becomes easier when we not only develop an understanding of the work and labor involved in the industrial production of animals as *food*, but we also begin to see these workers, both human and nonhuman, as individuals caught up in an economic and political framework that maintains their continued exploitation for profit. For researcher Kathryn Gillespie, author of *The Cow with Ear Tag #1389* (2018) and co-author of *Vulnerable Witness: The politics of grief in the field* (2019), "bearing witness" is important as it acknowledges that animals "are communicating their experience clearly" and such grieving with them "enables us to build a transformative politics of shared understanding, care, and nonviolent social relations" (2016: 584). It is a way, Gillespie argues, for us to be present in animals' intimate experience with structural violence (2016: 584).

To be clear, this is not an invitation to be a passive observer. Instead, we are invited to do something, "whether that is writing, speaking out, teaching about – to politicize hierarchies of power and inequality that work to oppress human and nonhuman others"

(Gillespie 2016: 585). Ceallaigh MacCath-Moran, folklorist and ethical vegan, has reflected on the complexities of her research, advocacy, and theorizing about human relationships with other animals. She offers her unique experience as someone who has volunteered in animal rescue, taken part in activism, such as bearing witness to the suffering of nonhuman animals on their way to slaughter, and lives a vegan lifestyle:

> I've read Francione, I've read Singer, I've read Donovan, and I want to use the words "animal welfare" versus "abolitionism" versus the "feminist caring ethic" that Donovan brought out, Josephine Donovan brought out, in some of her literature. I think that the part of me that is a wildlife rescuer is a welfarist, the part of me that is a dreamer and a hoper and someone who wants a vegan world is an abolitionist, and my soul is rooted in feminist caring. (Interview with Ceallaigh MacCath-Moran 2022)

It is important to contextualize the above quote by saying that in our interview, MacCath-Moran was very clear that she is not, in the general sense, an animal *welfarist*. The mention of animal *welfare* was not to self-describe as adhering to that framework but rather to describe her interaction with specific nonhuman animals in wildlife rescue and the care of her cats. In these contexts, meeting their specific needs and ensuring the well-being of the nonhuman animal in front of her is of utmost concern.

For example, MacCath-Moran mentioned that in caring for rescued birds of prey she may be required to feed them butcher's scraps, if the situation required it, while transporting them to a rehabilitation facility. This she used as an example of how "sometimes this is a place where all ethics are imperfect, where our ideals, our intellectual ideals go much, much farther than our practical capabilities can undertake. And so, when I am dealing with a situation that is directly in front of me, I look toward the welfare of the animal in front of me" (Interview with Ceallaigh MacCath-Moran 2022).

Resistance and the Beginning of a Just Transition

There are many courageous acts of resistance that have been undertaken by animal-based *food* production workers and their allies, and that hint at what a just transition away from industrial production of animals as *food* might actually look like. To begin, it is important to recognize that workers are already resisting in various ways. The most obvious form of resistance is the rates of turnover that exist in this industry (Blanchette 2021). Turnover rates in slaughterhouses in Canada and the United States range from 60 percent, all the way to 100 percent annually in slaughterhouses for chickens (Adams 1994: 82). Research has found that in Canadian *meat* processing plants, only one in ten workers were still on the job after their first year (Engebretson 2008: 236).

The stunning fact of the matter is that when we consider the question of how people continue to work in this industry, the answer is that they do not. Or, at the very least, they do not typically stay at the same facility for very long. Rather than simply thinking about this as quitting, anthropologist Alex Blanchette (2021) frames this turnover as refusal, and Carole McGranahan of the University of Colorado Boulder suggests that "[t]o refuse is to say no. But, no, it is not just that. To refuse can be generative and strategic, a deliberate move toward one thing, belief, practice, or community and away from another" (McGranahan 2016: 319). Blanchette, in response to the conditions of work for slaughterhouse workers in the United States during the COVID-19 pandemic, argues for an "anthropology of (non)work" (2021: 75), explaining that:

> I am trying to imagine stories and genres that decline to act in service of creating better conditions of work, or even entertaining that incremental workplace progress in these sorts of sites is viable. The ethnographic practice that this article desires is one that helps terminate these late industrial operations, as an end in itself – and learns from, with and for the people who cease working. (2021: 75)

As we saw in the preceding discussion, many slaughterhouse workers quit their jobs each year and this is important because it illustrates their refusal to perform dirty and dangerous work. As Blanchette argues, "Each and every year, the equivalent of the entire workforce quits. Even when there is not a virus sickening people for paltry wages, there are far more people who have refused kill floor work than there are people who actually do the work" (2021: 75). Saying "no," refusing to take part in that which one finds problematic, distasteful, or dangerous, can be clearly framed as a form of resistance to the structural violence workers experience and as an entry point to begin to think about what a regenerative food system might entail.

In addition to quitting, other worker-led strategies are also developed. Worker-created boycotts have been undertaken as a way of challenging unsafe conditions of labor. Some worker-led projects advocate for consumers to consider "a meatless May," and in some locations, slaughterhouse workers' children protest the unsafe working conditions of their parents out of fear for their parents' lives (Blanchette 2021: 74). The #Boycottmeat movement includes workers from slaughter facilities, and some worker-created strategies have advocated for boycotting *meat*, and even go so far as to suggest that plant-based diets be framed as a matter of public safety (Struthers Montford & Wotherspoon 2021: 80).

Another method of resistance to demoralizing and dangerous work advocates for a universal basic income (UBI) (Donaldson & Kymlicka 2020: 207–8). Advocates discuss the potential of such a progressive policy to strengthen workers' ability to refuse unsafe work and to facilitate a just transition away from intensive *animal agriculture*, and as an important counter to "bad/unfree/forced labour, to facilitate individuals' choice between different types of work, and to recognize all forms of work as equally deserving" (Blattner et al. 2021: 260). And, importantly, the potential of UBI is that it gives us the opportunity to consider the role of work in society, and what human and nonhuman labor might look like if it was separated from capitalist forms of production (Donaldson & Kymlicka 2020: 221–4; Blattner et al. 2021: 260).

While we have so far only examined resistance on the part of human labor under capitalism, it is important to recognize that nonhuman animals also resist (Colling 2021). Due to their subordinate position, animals' acts of resistance may not work out in their favor, but it is still important to recognize the role they may play in raising awareness, and in challenging the status quo. Sarat Colling (2021) writes of animals' resistance, pointing out the artificial borders and boundaries humans enact and the ways that countless nonhuman animals challenge the role they have been given by escaping, running, jumping, and hiding to avoid confinement and/ or death.

Carter and Charles write of nonhuman animal agency and how they act with intentionality and part of that agency is that they actively resist situations they do not like, and sometimes they are able to change the situation under which they live (Carter & Charles 2013: 322). Some of the stories of nonhuman animal resistance end with them being captured and shipped to the slaughterhouse anyway. Others meet a violent end at the hands of law enforcement agents (Pachirat 2011; Colling 2021).

Often citing public safety concerns, animals such as cows and pigs who escape from farms, slaughter facilities, or during transportation are sometimes killed by law enforcement when they elude capture. Often this is framed from a public safety perspective. There is often citizen outrage at such killings, expressed in news stories and on social media. It is interesting to consider this outpouring of concern when the killing of nonhuman animals for *food* is not generally questioned by most consumer-citizens.

Sometimes it turns out that these rebellious animals get to live out the remainder of their lives in retirement – compensation for their labor, if you will – at an animal sanctuary (Masson 2004: 152–3; Baur & Kevany 2020). In our interview with Gene Baur, co-founder and president of Farm Sanctuary, he asked us to imagine a more compassionate way of engaging with nonhuman animals, perhaps in contemplating what happens to nonhuman animals constructed as *surplus*. He says that we should consider their lives as having value outside of any economic worth.

> **Gene Baur, Co-founder and President of Farm Sanctuary:** It's kind of the same thing for a human being. It is agency, it is feeling able to express yourself fully, it is being in community with others who enrich your life, and I have seen all of these things at the sanctuary. Speaking to the importance of community, one of the animals I rescued a long time ago was Opie, who was a calf who had been born on a dairy farm and he was an unwanted male, so he was sent to the stockyard on the day he was born. He was still wet from the afterbirth. It was a freezing day in upstate New York, and he fell in an alleyway of the stockyard and was dying of hypothermia, and he was there in a crumpled mass, his eyes were sunken in, and he was practically comatose. I went to the stockyard manager and asked, "What's going on with this calf?" and he says, "Well, I have to bury him later today" and I said, "Well, what if I take him off your hands?" He said "Sure, go ahead". So, I brought this calf to a nearby vet, and she looked at him ... she said that he has very little chance of survival, it makes no economic

> sense, why are you wasting your time? I said, you know, to me this is not about economics, it's about trying to help somebody. She gave him intravenous fluids and I brought him back to the sanctuary and I took care of him overnight. And slowly the light started coming back to his eyes. He was able to lift his head. He was able to stand. He started nursing from a bottle. These are all good things. But he really wasn't thriving, so physically he was up but he wasn't thriving, and I was wondering what was going on here. And then I thought, he needs to be with his people, so I brought him out to the cow barn and the cows joined around him and he just perked up and that was what he needed. So, community is important too. (Interview with Gene Baur 2022)

Each retelling of stories of nonhuman animals resisting, or of human advocates going out of their way to rescue an animal from industrial *food* production, forces consumer-citizens to at least momentarily think about the value of these lives, as well as human responsibility to ensure that they have good lives.

Sometimes broad global events, such as the COVID-19 pandemic, bring attention to the inequalities, inequities, and resistance to the A-IC. As our interview with Fitzgerald revealed:

> One of the few positive things to come out of the pandemic [was that] there was increasing attention being paid to the work environment inside slaughterhouses because there was so much COVID transmission going on there. So, you know, a lot of people don't want to know what it's like inside of a slaughterhouse, but they had to [learn about it] if they were engaging with the regular mainstream news. (Interview with Amy Fitzgerald 2022)

As we have seen, workers in the industrial slaughter and *processing* industries suffered tremendously during the pandemic; they were more likely to be infected with the potentially deadly virus and were forced through law and/or circumstance to continue working even when their safety was at risk. The slaughterhouse corporations continued to make huge profits, despite the circumstances their frontline workers found themselves facing. Just like the public awareness raised when people hear about nonhuman animals' resistance, the pandemic also brought public attention to the suffering of nonhuman animals in this system.

News articles discussed the plight of so-called *farmed* animals during the pandemic, many of whom were killed when slaughterhouses shut down. It is ironic, though, that many people took notice of the killing of nonhuman animals only when they were being slaughtered due to a plant shutdown and were perceived as *surplus* to needs. It is interesting that many people seem to have no problem with the same species of animal being raised to be killed when the *product* becomes a *hamburger* or *chicken* sandwich. Similarly, some farmers, veterinarians, and slaughter plant workers discussed feeling blamed on social media, and by the public and peers, for *depopulation, food* waste and/or COVID-19 outbreaks (Polansek & Huffstutter 2020; Dryden & Rieger 2021; Baysinger & Kogan 2022), rather than people recognizing the structural dimensions that led to these outcomes.

Perhaps a good place to end this chapter is to contemplate a quote by Brandon Keim. He had been talking about both the rewards and challenges of his reporting on animal intelligence and promoting awareness of this body of scientific work. He discussed the importance of framing proposed social change and social justice-oriented changes as actually sharing much in common with deeply held values that so many of us already share, so that:

> A big challenge is just meeting other people where they are and trying to find the common ground or to nourish the spark of compassion that exists in their lives into something bigger and broader. And I don't say that like I'm some kind of wise person in this because, of course, the whole reason that I have the beliefs that I do now, the knowledge that I do now, is because people did this with me. And when I think my experience is probably pretty emblematic of the great many people where, you know, you grow up and there are some animals who are inside your circle and others who are outside the circle, but you still kind of respect them, and then just others who are completely invisible, and you don't think about what the boundaries are between those groups and how they're socially constructed and everything that goes into that. (Interview with Brandon Keim)

We already have the toolkit for creating a more just food system for both human and nonhuman animal laborers. We have learned about the structural and slow violence within this capitalist production system, but we have also heard about transparency, empathy, bearing witness, compassion, and resistance. As we move into chapter 4, we will begin to learn about what this intensive food system means for consumer-citizens, and how the concepts we have addressed in this chapter are helping to bring awareness to the unjust food system and how positive grassroots organizing is facilitating positive social change.

4 What If We Really Are What We Eat?

Challenging a Colonial-Capitalist Diet

> With colonial issues on a global scale, I also have to think about whether the food choices I am making are going to reinforce colonialism somewhere else. Am I benefiting here but helping oppress someone else? So, if you are going to decolonize, you cannot just decolonize locally, you have got to think about the strings that all the other decisions are attached to on a global scale.
> – Margaret Robinson

The most marginalized individuals and communities are not the ones most responsible for either the economic inequalities they face or the climate disruptions they are disproportionately having to endure. As was noted in chapter 1, Oxfam and other international agencies argue that global pandemics such as COVID-19, and increasingly frequent famines, fires, and floods, not to mention glacial melting caused by climate change, are only exacerbating these problems. They are also contributing to increasing levels of conflict around resources and land (Oxfam International 2022).

It is clear that addressing poverty, and the related issue of food insecurity, will require a multi-jurisdictional and intersectional approach. In addition, as climate scientists and scholars note, these must be combined with an emphasis on "social infrastructure" to ensure not only future justice and sustainability, but also to effectively and compassionately navigate the inevitable crises that will accompany more violent and unpredictable climate disruption (Klinenberg 2018; Jamail 2020; Bendell & Read 2021).

The current system of food production – industrialized and heavily reliant on nonhuman animals – fails to meet the needs of the global population. As Patel and Moore argue, "cheap food regimes" fail to feed the planet as evidenced by the ongoing "global persistence of diet-related ill health and malnutrition" (2017: 158). Below, we explore the true costs to consumer-citizens of "cheap food regimes" as well as how and why some communities are disproportionately suffering the consequences of what Vandana Shiva calls "cowboy capitalism." In addition to examining what a just food system needs to include, we begin by addressing the basic fact that we all have a fundamental right to food.

The Right to Food: Availability, Accessibility, and Adequacy

The Universal Declaration of Human Rights recognizes the human right to food (as well, we should note, the right to housing; United Nations 2009). In further covenants, such as the 1966 International Covenant on Economic, Social and Cultural Rights, protections were put in place to guarantee that *all humans* "have the right to adequate food and the right to be free from hunger" (United Nations 2010: 1). In addition to food being enshrined as a basic human right, it is also important to note that food must also be available, accessible, and adequate (p. 2).

"Availability" of food relates to both ensuring there is food to be found on the land or through the cultivation of the land with activities such as farming. People must also have access to food available for purchase. Secondly, "accessibility" relates to both "*economic* and *physical* access to food to be guaranteed" (United Nations 2010: 2). Therefore, food must be affordable, and people should not be forced to make "choices" over which necessity they will somehow try to do with less or without. This also ties into other areas of life, such as basic income guarantees or making sure that the minimum wage is sufficient to pay for food and other necessities of life. The United Nations makes clear that regardless of where someone lives, whether in a remote location or in an institutional setting such as a care facility for elderly people or a prison, access to food must be a basic guarantee. And thirdly, "adequacy" of food means that it must be nutritious and meet the needs of diverse humans, such as ensuring enough calories and nutrition to meet the needs of growing children or those pregnant and nursing, be free of harmful contaminants or substances (such as hormones, residues from pesticides, and contamination from industrial processes), and be culturally acceptable (United Nations 2010: 3).

In researching this chapter, we thought back to a resource we have often employed in our classrooms: asking students to think critically about when, where, what, whom, and whether we eat. Harper and Le Beau (2002) ask us to consider the adage, "Tell me what you eat, and I will tell you what you are," a claim made by Anthelme Brillot-Savarin, a French politician, gourmet, and the author of *Physiology of Taste* (Harper & Le Beau 2002: 4). Harper and Le Beau argue that asking "about food is like peeling back an onion. With each layer you peel away there are different meanings and problems" (Harper & Le Beau 2002: 3).

The onion analogy shows how when you start looking beneath the surface of food and its production, it is much more complicated than we might first imagine. As we begin to peel back the layers of industrial production of nonhuman animals as *food* and the implications for consumer-citizens, we see some of the key issues related to food production that place limitations on citizens' ability to eat or eat well, and also the additional consequences that extend beyond individuals and into their communities, cultures, and the broader global landscape.

In previous chapters we explored the historical process of "primitive accumulation" in capitalism that involved forcibly removing people from their lands, from the "commons." Commodification of land and resources, and therefore of food, has been and continues to be a critical challenge to the availability, accessibility, and adequacy of food. Food production became increasingly commodified as food became something that was bought with money (Harper & Le Beau 2002).

As medical anthropologist and farmer Joe Parish argues, "one of the big structural problems in our society is that we have gotten to the point where 1.7 percent of the population farms, and the other 98 percent don't. That's the whole country [of Canada]. It's a ridiculously small proportion because the farms are so big out west" (Interview with Joe Parish 2023). The worldwide decline in the number of farmers "parallels the inverse growth of multinational agribusinesses" and this growth in large-scale agriculture and centralized control is most pronounced in North American *livestock* production (Thu 2009: 15).

Wendell Berry argues that factory *farms* and intensive rearing and processing of nonhuman animals "raise issues of public health, of soil and water and air pollution, of the quality of human work, of the humane treatment of animals, of the proper ordering and conduct of agriculture, and of the longevity and healthfulness of food production" (2009a: 13). Therefore, it is important to think about the industrial production of nonhuman animals as *food* within a democracy framework given that consumer-citizens are left out of the decision-making process on an issue that has implications for health, well-being, and for the environment (Thu 2010; Gibbs 2017).

Under neoliberal globalization, there has been a concerted effort to decrease the areas of economic activity under democratic control (Gibbs 2017; Gibbs & Harris 2020: 146). When we are able to see our lives – past, present, and future – as bound up with the lives of other human beings, other species, and with the planet itself, we can begin to see that the effects of our individual actions extend far beyond our immediate surroundings and the political and legal boundaries of municipalities, regions, and the nation-state. What becomes clear is that the current ideas and practices of democracy, which do not extend to the economic sphere and don't take into account the realities of global economic relationships, are both an expression of structural violence and a barrier to creating a saner, more ecologically sustainable and compassionate world (Gibbs 2017).

It is important to acknowledge that access to quality and nourishing food is not just happenstance, but it is tied to inequality, and to structural violence. It is also essential that we recognize the intersections of oppression as the current food system, which not only fails to deliver the nutrient-rich food that facilitates healthy lives for many humans, but also leaches the soil of nutrients and treats animals as commodities rather than living beings (Interview with Vandana Shiva 2023).

According to agronomist, educator, and farmer Av Singh, under the current food system, "the mentality of what we're producing is that instead of being able to produce whole nutrient dense foods, we produce poor nutrient dense food, and then a whole bunch of byproducts that come from that to supplement a dysfunctional food system" (Interview with Av Singh 2022). Singh went on to say that within the current industrialized food system, most fail to recognize the essential interconnectedness between humans, other animals, and the land, and that:

> Not recognizing the fact of interconnectivity and that importance of, in my opinion, that whole garbage-in garbage-out mentality, which is our food

system, I think comes out of a colonial concept, an extractive mentality driven by that maximizing yield mentality as well. Those are all kinds of examples of a colonial philosophy that doesn't focus as much on quality, it doesn't focus as much on the integrity of a system, of reciprocity within that system. I think an extractive exploitative system is what we created, and we continue to develop. (Interview with Av Singh 2022)

This model of industrial agriculture, as Vandana Shiva also notes, has resulted in an approach that focuses on, "how do I extract the most out of a cow, out of the soil, out of a plant? Not even to meet human needs, but to feed the market and the commodity flow" (Interview with Vandana Shiva 2023).

Feeding the market is very different from feeding people. We often hear the claim that we simply do not have enough food to feed everyone on the planet, and that's why we have hunger and malnutrition. But according to organic farmer, George Naylor, the world produces enough food to feed 10 billion humans (2017: xix). Similarly, the United Nations reports that "Enough food is produced today to feed everyone on the planet, but hunger is on the rise in some parts of the world, and some 821 million people are considered to be 'chronically undernourished'" (United Nations 2019).

In reference to the fact that 828 million people went hungry in 2023, Action Against Hunger notes that, "After steadily declining for a decade, world hunger is on the rise, affecting nearly 10% of people globally. From 2019 to 2022, the number of undernourished people grew by as many as 150 million, a crisis driven largely by conflict, climate change, and the COVID-19 pandemic" (Action Against Hunger 2023). As Raj Patel points out, we have 800 million hungry people and one billion who are overweight (2009: 1). Rather than "over-" and "under"consumption of food being independent issues, they are each "symptoms of the same problem" as corporations produce food while influencing "how we eat, and how we think about food" (Patel 2009: 1).

The Issues with Overconsumption and Underconsumption

The issue, then, is not that there isn't enough production or enough food, but that food is commodified, and some people simply cannot afford to buy the basic necessities of life, or they live in areas where heathy food options are simply not available. Eric Holt-Giménez, Director of the Institute for Food and Development Policy, says that "most people in the world simply cannot afford to eat according to their values, [but] it is important for those who can to do so. But again, this doesn't change the basic commodity relations of value in the food system" (2017a: 71).

A 2011 report by World Watch Institute indicated great disparities in worldwide consumption rates, for both food and consumer goods. Inhabitants of North America and Western Europe make up 12 percent of the world's population, but they represent 60 percent of individual consumer spending (World Watch Institute 2016). Consumer-citizens in those regions consume on average "3 times as much grain, fish, and fresh water; 6 times as much meat; 10 times as much energy and timber ... as the resident of a poor country" (Bell & Ashwood 2016: 36).

The offshoot of this extractive profit-focused food economy is that we have countries, particularly in the Global North, in which most of the population is "overnourished," albeit often eating too much of particular types of food such as high fat, high sugar, highly processed foods containing refined sugars, carbohydrates, and fatty animal-based ingredients. This is often referred to as a "Western diet" (Cook 2004; Pollan 2008; Azzam 2021).

Diets higher in processed foods and fast foods are also often cheaper and more readily available in many locales, and consumers may look for things that will last in the cupboard or provide a quick meal from the freezer. This is especially true in lower-income areas. Due in part to government subsidies, processed and fast foods derived from nonhuman animals are often cheaper than whole grains, fruits, and vegetables (Simon 2013: 79–80). Why should this matter to us? Because as Robert Albritton states:

> Increased meatification of the global diet is a causal factor in starvation, because more and more of the world's grain production is being used to feed animals, and in the process, not only are many of the grain calories lost, the resulting meat is not generally affordable for the poor. In other words, meat is not a very efficient source of calories. Meat is the preferred food of hundreds of millions of people, but we need to consider the social and environmental costs associated with high levels of meat consumption. (Albritton 2009: 104–5)

Overconsumption of heavily processed and junk foods can have serious repercussions for health and longevity. According to the Canadian government, in recent self-reporting surveys, approximately

26.8 percent of adult Canadians reported being obese in 2018 (Statistics Canada 2019). There are regional disparities, and in our province of Nova Scotia, that number was closer to 33.7 percent, and on Cape Breton Island it was almost 42 percent of the adult population (Statistics Canada 2019). Obesity is just one aspect of this type of diet; diabetes, heart disease, some cancers, and other chronic conditions are also prevalent, as the groundbreaking research of Colin Campbell and Thomas Campbell made clear in *The China Study* (2006).

Furthermore, eating so-called junk food (high calorie, low nutrition), fast food (food that can be accessed almost immediately and on-the-go), and processed food (foods that tend to be less nutritious than their less processed counterparts) are not the only causes of obesity (Albritton 2009: 91–2). But as Professor Emeritus at York University Robert Albritton tells us, they are certainly "a major cause" (2009: 92–3) and "like sugar, fat plays a major role in making food taste better and hence has quasi-addictive qualities" (2009: 102).

Studies indicate that "some of the main sources of saturated fat in the US diet include dishes containing cheese, meat, poultry, and seafood" (Sterling & Bowen 2019). Albritton goes on to explain that "obesity is a concern because it correlates closely with the incidence of numerous chronic diseases including diabetes, heart disease and cancer" (Albritton 2009: 93). Referring to a study by University of Illinois at Chicago professor Jal Olshansky, Albritton explains that, when body fat is 30 per cent and above, life expectancy can on average be expected to decrease by up to ten years (Olshansky, as found in Albritton 2009: 93–4).

Deconstructing the "Ideal" Body

Several of our interviewees highlighted the importance of examining the destruction of traditional foodways under colonization and the racism embedded in privileging certain types of bodies over others, within both mainstream advertising and mass media, but also within more progressive change movements, such as some groups promoting veganism. In any discussion of weight, size, and diet, it is important to examine it from a systemic standpoint, such as examining the structural conditions within the current food system that make certain types of foods more readily available, widely advertised, and cheaper.

This focus on systemic issues rather than from an individual framing is also important because these food options are embedded in a colonial and capitalist framework. Within that, privileging of certain body types and sizes over others can be made to appear "normal" or "natural" and ignore that such conceptualizations also have ties to structural inequalities related to gender, class, disability/able-bodiedness, ethnic/racial inequality, and so on (see the works of Donovan & Adams 2007; Wrenn 2015; 2019; Greenebaum 2017; Taylor 2017; Harper 2020; Ko & Ko 2020).

Vegan sociologist Jessica Greenebaum discusses the importance of Critical Animal Studies (CAS) utilizing an intersectional framework and the importance of "black feminist critique of the universal experience of feminism in order to critique the campaigns and strategies of organizations like PETA that marginalize women, poor people, and people of color. The concept of the universal women's experience assumes race, class, and sexual neutrality, meaning that women are by default assumed to be white, middle-class, and heterosexual" (2017). In doing so, Greenebaum argues, an organization such as PETA "normalizes the white, thin, heterosexual body as the model of health and the normative image of vegans. This deters women who do not fit this image from learning about and participating in veganism, as they feel judged and excluded" (Greenebaum 2017).

A number of recent books has been written that counter this narrative and asks that we examine the privilege given to certain body types and the naturalization of a particular ideal body. As noted above, there is nothing *normal* or *natural* about such conceptualizations because they reflect the intersections of oppression that we have discussed throughout the book. These critiques are important in order to make plant-based and vegan diets and lifestyles more accessible to a wider range of citizens (see edited collections such as: *Veganism in An Oppressive World: A vegans of color community project*; *Sister Vegan: Black women speak on food, identity, health, and society*; *Apro-ism: Essays on pop culture, feminism, and black veganism from two sisters*).

Theorist and indie digital music producer, and founder of Black Vegans Rock, Aph Ko, and author Syl Ko argue that "[d]ismantling racism might require dismantling patterns of consumption, including food practices" (2020: 27). In our interview with folklorist Ceallaigh MacCath-Moran, she discussed a play she recently wrote, commissioned by the Odyssey Theatre in Ottawa for inclusion in a radio play series titled "The Other Path." She discussed how she modernized a fairytale on conceptualizations of beauty and in her adaptation "settled on fatness because people of size are often subjected to the public gaze in a very negative way, and I wanted to interrogate that in the play" (Interview with Ceallaigh MacCath-Moran 2022).

Any discussion of weight and size must be framed from an intersectional and culturally appropriate framework. Rather than looking at it in social isolation or as a "fact," a nuanced approach, as others have recommended, seems warranted where individuals and communities themselves create their own understandings that are culturally meaningful for them. Mi'kmaw scholar Margaret Robinson explained that in her research she has focused on the level of "cultural connections and cultural embeddedness that leads to different health outcomes. And so, it just struck me that culture keeps us well" (Interview with Margaret Robinson 2022).

Dr. A. Breeze Harper, Critical Race Feminist and Food Studies Scholar: So that comes up a lot, the fat phobia and how veganism is supposed to be equated with slim and fit, or slim as fit. I think that's another way in which the language needs to be decolonized. And these obsessions around, once again, purity, and purity being young, white, thin or skinny, and able bodied, so if you eat a vegan diet then it should equal this. So, language is more than just what we read, we have those images constantly being thrown at us. And I look into that when I talk to those who are doing vegan startups or vegan food companies when they're marketing and advertising that you need to understand the history of fat phobia as part of anti-blackness against black women.

There's this other, quite a bit of literature out there that talks about a lot of the fat phobia stems from a fusion of anti-blackness and anti-fatness, and the black more curvy body versus what a more civilized lady should look like, which is white and thinner and you know all this stuff. So just asking that people again really think about, you know, why do you think that a fit vegan body should be this way and who taught you that? And you know, how accurate and valid is that? And how much disgust do you have come from certain bodies, how much of that actually comes from, once again, like a white supremacist

> caste system that tells you certain bodies are more ethical than others?
>
> Because whiteness isn't just phenotypes, it's a bunch of ideas as well that end up creating a certain type of civilized and ethical whiteness that has changed over time. So those are the things that we discuss and unpack as people are trying to market a new product or find images or create a video for their new product or their new line, to kind of tackle that. (Interview with Dr. A. Breeze Harper 2022)

Building Community and Social Infrastructure

It would be easy to blame many in the Global North for their overconsumption of food and consumer goods, but this increase in consumption came about because of a progressive and relentless shift from citizens' consuming, to 'consumer' becoming a primary identity for many, often surpassing that of citizen (Schor 2011; Gibbs 2017). The individualistic culture promoted by neoliberalism and the related erosion in, and of, community, coupled with fewer opportunities for civic and meaningful social interaction in community, means that citizens of the Global North have been conditioned to look out for themselves. Additionally, as previously mentioned, many of the links between consumption and its harmful effects (on the environment, humans, and other species) are not immediately visible to consumers.

Sociologists Krishan Kumar and Ekaterina Makarova note the importance of recognizing "the decline of socialism, the attack of all forms of public organization – from health to transport – as wasteful and inefficient, the relentless privatization of cultural and educational institutions" (Kumar & Makarova 2008: 325). And yet, as research tells us, our connection to others, in community, is what insulates us from many hardships, and communities with more opportunities for social engagement fare better when faced with crisis, such as extreme weather events or pandemics (Klinenberg 2018; Li et al. 2022).

It is essential we recognize that citizens do not always have the opportunity to avoid the push towards individualistic tendencies as opposed to collective ones (Harris 2018: 7). According to sociologist Janet Lorenzen, citizens are subject to "consumer lock-in" when they do not have access to public services such as public transit and therefore must instead utilize individualistic modes of transportation such as owning their own car. She argues that "consumer lock-in" is most significant for those with "limited resources," because in these cases people are "locked into particular consumption patterns," and that this important fact makes clear the need to challenge "structural inequalities" in all communities. Doing so would provide ways to support the most vulnerable, create meaningful modes of sharing, invest in spaces for people to connect, and all of these would help us tackle food insecurity, emissions, and climate issues collectively (Lorenzen 2014: 1072).

Public investment, or lack thereof, in social infrastructure relates to many of the themes we have already addressed. Neoliberalism, for instance, gutted many communities through the loss of manufacturing jobs, often well-paid unionized positions. Many communities have also lost local grocery stores and coffee shops, as well as experiencing an underinvestment in social infrastructure like parks, libraries, sidewalks, public schools, community centres, and so on (Klinenberg 2018). Social infrastructure is like a bridge that connects people to each other and ensures that they have a safety net. It can include neighborhood parks, walking trails, playgrounds, farmers' markets, the local YMCA, libraries, etc. The best types of social infrastructure do not focus on efficiency, but rather help facilitate time with others that helps build "cooperation and trust" (Klinenberg 2018: 18).

At the beginning of this book, we discussed some of the challenges our community recently faced with Hurricane Fiona. It was truly a stressful and taxing time, as people struggled with the loss of their homes, personal property, and food spoilage, the lack of power and water, and the uncertainty about when things would start to get back to normal. What happens when the hard infrastructure (such as roads, sewage, water, power, and food delivery to grocery stores) collapses during an extreme weather event? According to Klinenberg:

> When the power goes out, most businesses, health care providers, and schools cannot operate, and many transit and communication networks stop running too. Breakdowns in the fuel supply can be even more consequential, since oil generates most of our heat and gas powers the trucks that deliver nearly all the food and medications consumed in large cities and suburbs as well as the automobiles that most people depend on to travel ... And, as most policy makers and engineers see it, when hard

infrastructure fails, as it did in the great Chicago heat wave, it's the softer, social infrastructure that determines our fate. (Klinenberg 2018: 15)

When social infrastructure is neglected and degraded, it may not be as dramatic as a washed-out road or fallen power line, but "the consequences are unmistakable. People reduce the time they spend in public settings and hunker down in their safe houses. Social networks weaken ... Distrust rises and civic participation wanes" (2018: 21). One of the things that made Hurricane Fiona survivable was the social infrastructure in this tight-knit community. There were so many examples of people helping each other and the established social infrastructure expanding their role in community to help even more people during a difficult time.

The local YMCA, for instance, opened to everyone in the community, offering a warm, dry space to charge electronic devices, use the internet, eat a free meal or snack, and to have a hot shower. In a similar vein, schools, firehalls, and community centers offered food, internet, and other supports. Due to the damage to power lines, often from fallen trees, many communities throughout the Canadian Atlantic provinces lost power, and in many jurisdictions, often for extended periods of time (weeks not days). The First Nations community of Eskasoni experienced a prolonged power outage and "Potlotek First Nation sent 22 generators, and Wagmatcook and We'koqma'q First Nations ... sent food, water and gas" (Baker 2022).

There were also multiple examples of international students from Cape Breton University opening impromptu kitchens in their neighborhoods and offering free meals to their neighbors (Jala 2022). These local examples demonstrate the diverse ways that social infrastructure can insulate and protect us from extreme weather and other natural disasters, while bringing people together in caring, compassionate interactions that help to safeguard democratic principles of trust, engagement, and civic participation.

From Food Insecurity to Food Justice

At a societal level, we struggle with so-called "lifestyle diseases" – heart disease, diabetes, hypertension, to name a few. As we have learned, these diseases are not uniformly distributed throughout society, as poor communities, as well as Indigenous, Black, and other racialized communities, tend to suffer the effects of this "lifestyle" most often (Sterling & Bowen 2019).

Facilitator, Black vegan activist, and food justice researcher Starr Carrington writes that in the United States "food access in predominantly white, middle-income areas, on average, far exceeds food access to predominantly Black, low-income areas, which represents a racial divide within our food system" (2018: 176). Within poor communities, access to whole food, that which is natural and not processed, is often limited and citizens are forced to buy food from convenience stores, liquor stores, and fast-food restaurants.

For years these areas have been labeled "food deserts," but more recently the term "food apartheid" has been used because it conjures the structural violence, such as "the corporate and political polices that led to the current landscape: racism" (Robinson-Jacobs 2021). Activist Karen Washington coined the term "food apartheid" because:

> Oftentimes, people use the words "food desert" to describe low-income communities who have limited access to food. In fact, we do have access to food – cheap, subsidized, processed food. The word "desert" also makes us think of an empty, absolutely desolate place. But there is so much life, vibrancy, and potential in these communities. I coined the term "food apartheid" to ask us to look at the root causes of inequity in our food system on the basis of race, class, and geography. Let's face it: healthy, fresh food is accessible in wealthy neighborhoods while unhealthy food abounds in poor neighborhoods. "Food apartheid" underscores that this is the result of decades of discriminatory planning and policy decisions. It begs the question: What are the social inequities that you see, and what are you doing to address them? (Brones 2018)

This discrepancy in food access, and its connection to policy, has led the Director of Healthy Food Access Programs at The Food Trust, Brian Land, to prefer the term "food redlining" to describe the deliberate practices at work in unequal food options in predominantly Black and poor communities in the United States. The term "redlining" was traditionally used to describe the systemic denial of residents in certain neighborhoods trying to access mortgages or loans.

It is defined by the Fair Housing Center of Greater Boston (n.d.) as "the practice of denying or limiting financial services to certain neighborhoods based on racial or ethnic composition without regard to the residents' qualifications or creditworthiness. The term 'redlining' refers to the practice of using a red line on a map to delineate the area where financial institutions would not invest."

This practice led to the "legalization and institutionalization of racism and segregation" (Fair Housing Center of Greater Boston n.d.), just as the food choices in poor and predominantly Black communities are based on the market forces and public policy decisions that favor predominantly white communities over their predominantly Black counterparts (Lang, as found in Robinson-Jacobs 2021). All these issues point to the structural violence underpinning contemporary food systems.

We asked Dr. A. Breeze Harper, Critical Race Feminist and Food Studies Scholar, about the importance of the language we use to describe such systemic food insecurity. She began her response by giving credit for the term "food apartheid" to Karen Washington, and then explained:

> It is important to – once again, we are talking about decolonizing language – use language that explains the history of oppression and the outcomes that you experience today. So, to call a space that has horrible access to food, healthier foods, a desert, it erases the reality that it was orchestrated to be that way because a desert is a naturally, usually a naturally, occurring ecosystem, and it's an ecosystem of its own, it's not like it's void of what animals and plants need that evolved living there. But when we say apartheid, that is orchestrated. Apartheid exists so you could subjugate people on purpose, so I think that is why it is important to use the term apartheid when we are talking about the issue of food. (Interview with Dr. A. Breeze Harper 2022)

The dietary options available in this unjust system stand in sharp contrast to the traditional diet of Blacks in the United States. Samara Sterling and Shelly-Ann Bowen point out that the "'old ways' of eating consisted mainly of green leafy vegetables, sweet potatoes, fruits, beans, peanuts, coconuts, homemade sauces, herbs, and spices," therefore the traditional diets of "Blacks were historically predominantly plant-based" (Sterling & Bowen 2019). They argue that eating more plant-based options could not only save consumers money, but with the added benefit that such diets are known to reduce several diseases such as cardiovascular disease and type 2 diabetes (Sterling & Bowen 2019). Research on the affordability of predominantly plant-based diets has had contradictory findings

(Guillemette & Cranfield 2012; Lusk & Norwood 2016). But we do know that the costs of plant-based whole foods could be reduced with a redistribution of subsidies that now support feed for *livestock* production and other industries involved in producing nonhuman animals as *food*.

In addition to issues of access to healthy and affordable food, many people are simply too busy and too tired to contemplate cooking after a long workday, and supermarkets and fast-food restaurants have stepped in with convenient, processed, and heavily packaged premade meals. As sociologist George Ritzer points out, this "progressive rationalization has threatened … the health, and perhaps lives, of people" (2015: 129). One fast-food meal, for instance, may contain an entire daily recommendation of "fat, cholesterol, salt, and sugar" (Ritzer 2015: 129). As we discussed earlier in this chapter, the result is epidemic numbers of diet-related illnesses, as well as mass advertising that promotes a lifetime of unhealthy eating habits for children (Ritzer 2015: 130).

These disturbing realities and the related inequities in food options make it clear that some form of food justice is needed. But what exactly is the meaning of the term "food justice"? According to Harper:

> People need to understand what we mean by justice and that it is not something that is random, so when you say justice, what type of justice? What are you talking about when you merge it with food? It is important to note the limits of justice in a capitalist system. Capitalism has not proven to me that [it] is the way to go if we are going to create a regenerative food system. It's important to problematize that and think about reparations and land. How can we talk about that when most of us in North America are on stolen unceded land. (Interview with Dr. A. Breeze Harper 2022)

Margaret Robinson, Mi'kmaw Scholar, Canada Research Chair in Reconciliation, Gender, and Identity, and Member of Lennox Island First Nation: Lots of work has been done on particular colonial diets that have increased diabetes, for instance, in First Nations. And some of the work has looked at dietary changes, but they've mostly been at the individual level looking at "how can I as a Mi'kmaw woman eat differently so I don't get diabetes or so to treat my diabetes?" They rarely looked at the

> systemic relations of people and food, that look at "Where are the colonial determinants of health here?" In terms of the ways they're shaping people's access to food, access to traditional food, the kinds of things they can afford, how much things cost where they live. What they can afford is really one of the biggest determinants.
>
> But I didn't see a lot of Indigenous health work looking at food in relation to impoverishment. So, I started thinking about it in relation to traditional food economies and health outcomes resulting from their destruction. I thought maybe the issue isn't just food and movement of food, but also access to culture and depth of enculturation. And that got reinforced by work done by some folks out West (Oster et al. 2014) who looked at Indigenous language fluency as a proxy for cultural embeddedness and found that people who spoke their Indigenous language had lower diabetes risk.
>
> We know that learning Mi'kmaw is not going to make me less diabetic, but there is something about the cultural connections and cultural embeddedness that leads to different health outcomes. Culture keeps us well. And this is a theme that you see reflected in other groups who are similarly affected by colonial structures. (Interview with Margaret Robinson 2022)

The discussion we had at the beginning of this chapter, about food being a basic human right, relates directly to food justice (Rodriguez 2018a: 8). Aric McBay, author, farmer, and organizer, spoke to the intersecting themes involved in food justice when he said:

> For me, climate justice remains the most urgent issue that I work on. I work on lots of things. I work on prisoner justice, anti-poverty, and all these other causes. But if we're not able to make dramatic action on climate change in the immediate future, then the prospects of all of our movements are pretty bad. Our opportunities are going to be very limited. So, for me, climate justice remains kind of the centerpiece, the hub. But so is food, right? Because food and land connect somehow with basically every other justice concern, with every equity concern. (Interview with Aric McBay 2023)

For interview participants, even though their particular areas of expertise in research and/or activism were in unique fields (animal protection, agroecology, decolonization, climate protection, worker rights, and so on), they all recognized the connection to food and food justice.

We asked Av Singh about the connection between his engagement with food and the connection with social justice, and he stated that there will be no real change "until you address the fact that we are on occupied land, and the fact that women have been traditionally excluded from our current agricultural system, and persons of color are not visible on the land. We can sequester more carbon but at whose expense, and whose continued expense?" (Interview with Av Singh 2022).

Ensuring that all people, no matter where they live, or their gender, ethnicity, socioeconomic situation, and so on, can eat well, is ensuring food justice. Food security is the understanding that a group should be able to feed themselves and they should also be able to decide how they feed themselves. And, finally, food insecurity relates to "household-level economic and social conditions of limited or uncertain access to adequate food" (Rodriguez 2018a: 8). Vandana Shiva pointed out that food justice is not only connected to humans, but also related to all beings and the soil in the "web of life." According to Shiva:

> The issue of what's called our food sovereignty, with what I call *Swaraj*, is the ability to provide for yourself, you know, as community. So, growing your food is part of food justice and food justice for eaters means having adequate food, having food that doesn't kill you because food is now the biggest killer; having food that is not poison, ultra-processed, etc. And the right to grow your food and the right to eat cannot exist independent of each other. And therefore, food is a community. (Interview with Vandana Shiva 2023)

Saryta Rodriguez argues that there is a strong connection between vegan/plant-based diets and food justice communities, and that providing access to more whole-food, plant-based options would go a long way to easing food insecurity worldwide:

> I have come to understand, as a food justice worker first, and then later as a vegan, that it is in fact not only *useful* to discuss animal agriculture when attempting to promote food justice worldwide; it is *essential*. To attempt to solve the problem of a lack of food justice without taking into consideration the detrimental impacts of animal agriculture on us, others, and the planet is like trying to understand how cancer spreads without knowing anything about cell growth. (Rodriguez 2018b: 86)

Mi'kmaw scholar Margaret Robinson notes the complexity of trying to solve the issues with food insecurity because, "in addition to being a colonial diet, part of the problem is that it is a capitalist diet" (Interview with Margaret Robinson). Vegan sociologist Jessica Greenebaum echoed this:

> We cannot separate ourselves from this capitalist system as much as we want to. I think critiques of capitalism are important, but we must be realistic, right? There must be some combination of these big global food systems, but also micro local systems. How do you empower small communities and people to feed themselves? So, I really love the idea of urban farming. (Interview with Jessica Greenebaum 2022)

Both Robinson and Greenebaum recognize that "*Capitalism* is the silent ingredient in our food" (Holt-Giménez 2017a: 233). Identifying the issues with food being entwined with capitalism, and trying to think through strategic grassroots ways of building resilience, was in fact a common theme throughout many of the interviews. This clearly connects to the earlier discussions of the intersections of oppression and how the global economic system, based on the capitalist logic of extraction and the constant need for more, is reinforced by colonial tactics of land appropriation, environmental destruction through *resource* depletion, and even violence and genocide in the creation, production, and transportation of various goods, including food.

Julia Feliz Brueck argues at the beginning of *Veganism in an Oppressive World: A vegans of color community project* that "All social justice movements are interconnected because we humans exist within a system that relies on inequality. As vegans of color, we fight for nonhuman animal rights, yet we also have to fight for our own rights in a world based on white supremacy and systemic oppression" (2017: iii). In expanding on the idea of various forms of justice as interconnected, we opened this chapter with Margaret Robinson's elaboration on how we need to think about food justice both at an individual and systemic level.

How do we forge ahead, and try to position ourselves as local citizen-consumers, but within this global food system? In Singer and Mason's *The Ethics of What We Eat* (2007), we are reminded that the ethical principle of "needs" means that just because we like the taste of a particular food item, or the convenience of it, that is not sufficient reason to continue with such production and consumption, if doing so enacts negative externalities (spillover costs) onto workers, residents, consumers, and other animals used in food production, or if it leads to the destruction of natural habitats endangering free-living nonhuman animals (Singer & Mason 2007: 270–1).

Thinking about real "needs" rather than individual "wants" opens the possibility of broadening the conceptualization of not only individual responsibility but the possibility of championing systemic solutions to fight for food justice. And it is also true that cultural practices can and do shift over time as communities adapt to the changing conditions of their environment.

Amy Martin, Research and Writing Fellow at Food Tank, a US nonprofit organization promoting safe, healthy, and nourishing eating, discusses several ways to fight for food justice. These include land access, saving seeds, defending worker rights, and working towards better food education (Martin n.d.). Each of these options speaks to food justice as they imply building the capacity for farm workers, farmers, and consumer-citizens to make decisions about the food system, rather than having those decisions made for them.

Such a shift in the food system might enable us to plant the seeds of some form of food democracy. To get to a place where we can start to examine ways in which this could happen, and echoing the recommendations of Martin, we are now going to discuss some of the additional costs – individual, societal, and global – of the current industrialized food system, so that we can see both where we currently are and where we could possibly go in the future.

Zoonotic Diseases: How We Treat Other Animals Comes Back to Haunt Us

Adequacy of food also relates to food safety because, in addition to people having the opportunity to consume enough calories from food that is culturally appropriate, the United Nations also highlights that food must be free from harmful contaminants or substances (United Nations 2010: 3). One major area of concern relating to contaminants is zoonotic diseases, which are infectious diseases that are transmitted from nonhuman animals to humans, or vice versa. Within this section we are going to examine several types of zoonotic diseases: foodborne illnesses, prion diseases, and influenza.

Foodborne and Prion Diseases

Foodborne illnesses are not all zoonotic, but they are all contracted through the alimentary tract (i.e., anything you eat or drink). The egg salad that has been left outside too long at a family picnic (foodborne but not usually zoonotic), the poultry farmer infected with bird flu (zoonotic but not foodborne), and the salad greens from the grocery store that were contaminated with *E. coli* through unsanitary handling and/or growing practices (zoonotic and foodborne).

Unsafe food can cause a wide variety of illnesses, which can lead to discomfort for many sufferers, but also disability and death. This threat of foodborne diseases is most pronounced for children and people with compromised immune systems, including the elderly (World Health Organization 2015a; Government of Canada 2021a).

Rather than being an isolated issue, divorced from the other major themes discussed in this book, relationships with the land and other animals are connected to the potential risks of certain zoonotic diseases. These may infect consumers and people who work with nonhuman animals in *farming* and in *meat* and *chicken* slaughter and *processing*. It is clearly important to understand how industrial practices may influence the prevalence of diseases. For instance, foodborne diseases are thought to be closely connected with the way we raise and transport other animals under industrial methods (Greger 2007; Guthman et al. 2014) and, as a result, "food-producing animals are the major reservoirs for many foodborne pathogens" (Heredia & García 2018).

Even though the majority of people infected with foodborne illnesses do recover, the World Health Organization is clear that we should not become complacent because globally approximately 600 million people become sick each year, and of those approximately 420,000 die (World Health Organization 2015a; n.d.; Lee & Yoon 2021: 1). Children under five years of age are most at risk; it is estimated that globally 125,000 children die annually from foodborne illnesses (World Health Organization 2015a).

While the World Health Organization identifies a number of factors that contribute to foodborne diseases – contamination of water, problems with food storage, regulations being poorly enforced – they also acknowledge that "intensive animal husbandry practices are put in place to maximize production, resulting in the increased prevalence of pathogens in flocks and herds" (World Health Organization

2015a: 3). This is not to say that these types of diseases did not exist in previous generations, or even in places that may still have a reliance on small-scale agriculture. However, there are several issues related to food safety and the conditions under which animals are currently raised that make the associated risks much more pronounced.

With the intensive production of animals as *food*, some health risks have become commonplace, such as the foodborne diseases *Salmonella* and *E. coli* (Sorenson 2010). *Salmonella* is a common "public health concern across all regions of the world, in high- and low-income countries alike," whereas some other foodborne diseases, such as *E. coli*, "are much more common to low-income countries" (World Health Organization 2015b).

The increased use of antibiotics in the industrial production of animals as *food* has also created the concurrent issue of antibiotic-resistant strains of infections in both humans and other animals. It is believed that "livestock are a primary source of antibiotic-resistant *Salmonella*, *Campylobacter* and *Escherichia coli* strains that are pathogenic to humans" (Rohr et al. 2019: 450). Antibiotic-resistant genes from raising fishes in confined spaces (*aquaculture*) are also thought to be "transferred to human systems and these pathogens have caused outbreaks" (Rohr et al. 2019: 450).

E. coli, for instance, is regularly present in both human and nonhuman animals' intestinal tracts. We only get sick when our own strains get out-competed by foreign (to our bodies) strains (American Society for Microbiology 2011). Meanwhile, *Salmonella* is regularly present in nonhuman animals who are *farmed*. Therefore, any bedding, food, and water that is used with one group of animals should be kept far away from other animals and humans (College of Veterinary Medicine 2016). Again, the scenario from the discussion of *E. coli* applies here as well.

It is more important to keep *food* products made from nonhuman animals clean than to dose nonhuman animals with antibiotics, and the WHO strongly recommends an "overall reduction in use of all classes of medically important antimicrobials in food-producing animals" (World Health Organization 2017: xii). The former aims to limit human contact with the bacteria responsible for the illness but allows the normal bacterial numbers that exist in the immune system of the animals to exist. The latter aims to limit the numbers of bacteria in *farmed* species so that housing conditions can continue to accommodate more nonhuman animals per square foot, without causing disease to spread easily. This creates the secondary problem of imposing unnecessary selective pressure on the bacteria to become the strongest and most resistant strains to survive.

Sara Shields from the Humane Society International and Michael Greger from the Humane Society of the United States argue: "In the egg industry, caging chickens [in battery cages] has been definitively tied to increased food safety risk" (Shields & Greger 2013: 393). Health Canada's website contains a lengthy list of recommendations for safe food handling and preparation of *chicken, eggs, hamburger, shellfish*, as well as some fruits and vegetables. Recommendations for egg consumption include warnings, such as never eating cookie dough made with raw eggs, due to the potential for *Salmonella* poisoning and ensuring that eggs are cooked to an internal temperature of at least 74 degrees Celsius to kill any potentially harmful bacteria. So dangerous is the potential for contamination that Health Canada recommends immediately cleaning cooking utensils with soapy water and then cleaning up with a mild bleach solution (Health Canada). Meanwhile, *E. coli* can make its way into the *food* supply by contamination of *meat* with manure, or if the contents of an animal's digestive system are spilled during slaughter. Both become harder to avoid in the fast pace of modern slaughterhouses (Fitzgerald 2015: 96).

While some countries may routinely test the "microbial load of table eggs" prior to them being sold in grocery stores, such practices may not be followed in all countries (Okorie-Kanu et al. 2016). In one study from Nigeria, it was found that the sample eggs were often contaminated with *Salmonella* and/or *E. coli* and that the strains found were resistant to several low-cost antibiotics that are commonly used in the Global South (Okorie-Kanu et al. 2016). Some suggestions to try and limit the spread of *E. coli* include cleaning chicken barns thoroughly, proper ventilation on farms, and safe food preparation, such as cooking at a hot enough temperature (Kabir 2010).

Salmonella remains "a serious public health concern because most of the strains of *Salmonella* are potentially pathogenic to humans and other animals. Avian salmonellosis, for instance, can pose a health risk to people if exposed. Symptoms appear like food poisoning, such as diarrhea and acute gastroenteritis" (Kabir 2010). For *Salmonella*, precautions include safe food handling and preparation, wrapping *meat* in plastic bags to avoid cross-contamination, and cooking at high enough temperatures to kill harmful bacteria (Kabir 2010).

For warnings related to leafy greens, the issue is not the greens themselves but rather that they may have been contaminated with *Salmonella*, *E. coli*, or one of the many other pathogens from the growing process (e.g., human waste being used as fertilizer). They may also have been irrigated or washed with contaminated water, have had contact with nonhuman animals, or used improperly composted manure in the growing of the crops (Interview with Joe Parish; Jung et al. 2014; Government of Canada 2016; Sofos 2008). Such produce may also be contaminated during the harvesting process, in storage, or during transport. Finally, Health Canada warns that leafy greens can also become contaminated "with harmful bacteria from raw meat, poultry, or seafood," in retail outlets, in someone's refrigerator, or during cross-contamination during food preparation (Health Canada n.d.).

Although many foodborne illnesses are bacterial, like *E. coli* and *Salmonella*, there are several that are viral (e.g., Hepatitis A) and prion-based (e.g., bovine spongiform encephalopathy [BSE], a.k.a. "mad-cow disease"). Prion-based diseases like Creutzfeldt–Jakob disease (CJD) in humans or BSE in cows are some of the most frightening diseases because they are difficult to keep track of, and there is no way to test a live cow for the disease (US Food & Drug Administration 2020). Once infected, BSE is always fatal because these diseases attack the central nervous system (Government of Canada 2022; Centers for Disease Control and Prevention 2021b). Prion-based diseases are essentially a misfolded protein, so they are proving challenging to design an effective test for, and researchers do not yet have a clear sense of how they originate (Peden et al. 2021).

While prion diseases are not always tied to intensive farming practices (e.g., chronic wasting disease in deer populations), they can cross-infect related species, such as elk and deer *farmed* for food (Centers for Disease Control and Prevention 2021a; Kurt & Sigurdson 2016: 83). One known mechanism of infection between cows has been the feeding of nonhuman animal remains to other *livestock*. For this reason, since 1997, Canada has legislation that bans feeding cows, sheep, and goats "mammalian-derived proteins" (e.g., feed made from other mammals) (Government of Canada 2022).

Products made from BSE-infected cows can infect humans who ingest them. It can further spread between humans through exposure to blood products, such as through a blood transfusion. Not all prion-based diseases can cross from nonhuman animals to humans. For instance, scrapie, a disease that affects sheep, has not been found to infect humans (Concepcion et al. 2005: 919). It is important to note that to date zoonotic prion-based diseases are not common in human populations but the lack of understanding of how they originate, and the deadliness of them, is certainly a cause for concern.

Although many zoonotic diseases existed prior to the industrial production of animals as *food*, research suggests that there are two main differences today: (1) the unprecedented rate of emergence of infectious diseases, and (2) the global projections that there will continue to be a sharp increase in food demand as the human population continues to grow (Rohr et al. 2019: 451). It is also believed that as there continues to be more global demand for *meat*, and as both people and nonhuman animals live in higher densities, the potential for infectious diseases will also increase (Rohr et al. 2019: 451). This also speaks to the increase in the farming of nonhuman animals that were not previously domesticated, such as deer, and the potential for introducing diseases from those nonhuman animals into domesticated animals and human populations (FAO, OIE & WHO 2010: 14).

In the 2008 report, *Putting Meat on the Table: Industrial farm animal production in America,* the Pew Research Center discussed some of the changes that have happened in the way in which nonhuman animals are *farmed*, and what that means for the lives of those animals, in addition to the potential risks of the introduction of diseases to human populations:

> Fifty years ago, a US farmer who raised pigs or chickens might be exposed to several dozen animals for less than an hour a day. Today's confinement facility worker is often exposed to thousands of pigs or tens of thousands of chickens for eight or more hours each day. And whereas sick or dying pigs might have been a relatively rare exposure event 50 years ago, today's agricultural workers care for sick or dying animals daily in their routine care of much larger herds and flocks. This prolonged contact with livestock, both healthy and ill, increases agricultural workers' risks of infection with zoonotic pathogens. (Pew Commission on Industrial Farm Animal Production 2008: 13)

This density, coupled with the unnatural housing conditions, is a major barrier to raising healthy nonhuman animals. Nonhuman animals and farm workers both suffer, and it creates the potential for ill health in both. What is needed is a more holistic approach that fully examines the connections between the industrial raising of nonhuman animals as *food*, human health, and the ways in which the improved quality of life for nonhuman animals (called *welfare* by some) may also help reduce such disease in both nonhuman animals and humans (Pew Commission on Industrial Farm Animal Production 2008).

The current approach that focuses on one issue at a time (e.g., an outbreak of a particular zoonotic disease) is obviously not working. Instead, we need a systems-based approach that examines intensive production of animals and asks why these problems keep emerging and how we can do better.

Influenza

The COVID-19 pandemic raised awareness of the ongoing global health threats posed by intensive confinement in the raising and slaughtering of nonhuman animals for *food*, as well as the repercussions of encroachment on nature and the ecosystems of other animals more generally (Rohr et al. 2019: 450–1). Part of creating a better post-COVID world is taking seriously the relationship with other animals and the connection between how they are treated in intensive *food* production, the toll we take on the natural world through encroachment and habitat destruction, and the potential for another serious zoonotic pandemic (Quammen 2012; Fitzgerald 2015: 98–9; Greger 2020).

Zoonotic influenza refers to those infections able to jump from nonhuman animals to humans (Quammen: 2012: 20–1). This is not a direct jump from the original host species to humans, but rather the virus spills over from the original host to an intermediary animal, before sometimes infecting humans (Quammen 2012; Greger 2020). Typically, the virus is not harmful to the original host animal but can become more dangerous when it infects another animal species, including humans (MacKenzie & Jeggo 2013: 170). According to the World Wildlife Fund (WWF), such jumps can be considered spillover. Alarmingly they also tell us that they are becoming more common as "three out of four new diseases are zoonotic" (World Wildlife Fund 2020).

The crux of the problem is not the animals themselves, "but the way we treat them that is the true cause of the pandemic" (Blattner 2020: 41). Increased industrial production of nonhuman animals facilitates increased probability of human diseases from animals

(Rohr et al. 2019: 451; Greger 2021), as we saw above in the discussion of zoonotic foodborne diseases. Some of the best-known influenzas today are those known as 'swine flu' and 'bird flu' (Fitzgerald 2015: 98–9). It is important to understand the conditions that make spillover more likely (Quammen 2012: 21).

Crowding of nonhuman animals contributes to stress, and stressed animals may be more susceptible to infection, and because of proximity in intensive agriculture, infected animals will come in close contact with other animals who may in turn be infected (Pew Commission on Industrial Farm Animal Production 2008: 13). In addition to the stress from crowding, other animals raised in intensive agriculture also have other modifications and life events that can increase their levels of stress; they are often weaned early, live in artificial settings, have been bred for uniformity of traits, and so on (Greger 2020: 99–103). In fact, this lack of genetic variation is perfect from a "virus's point of view" because once it is able to "infect one bird or beast, it can spread without encountering any genetic variants that might otherwise slow the bug down" (Montenegro de Wit 2021: 103).

It appears that the zoonotic disease drivers we currently face are all exacerbated by humans, because they result from human activities. This does not mean that all zoonotic diseases carry the same level of threat or potential to become a human pandemic, as they vary in their transmission and spillover potential. But the COVID-19 pandemic may be a dress rehearsal for the potential devastation to human life that might result if we were to see a widespread outbreak from a more highly contagious flu variant, such as H5N1 (Greger 2007).

H5N1 is better known as avian influenza or "bird flu." Already decimating some animal populations, this influenza can be of a lower pathogenic loading, or it can be highly pathogenic in nature. The US Geological Survey points out that "the designation of low or highly pathogenic avian influenza refers to the potential for these viruses to kill chickens. The designation of 'low pathogenic' or 'highly pathogenic' does not refer to how infectious the viruses may be to humans, other mammals, or other species of birds" (US Geological Survey n.d.).

Detection of the highly pathogenic avian influenza has been found throughout many countries, and here in Nova Scotia it has been found in commercial and backyard flocks of birds (such as chickens), in wild birds (such as crows and bald eagles), and in wild mammals (such as foxes, racoons, and seals) (US Department of Agriculture, Animal and Plant Health Inspection Service 2023; National Wildlife Health Center 2022)

At this time, H5N1, even the highly pathogenic avian influenza (HPAI) version, is capable but not very adept at jumping from nonhuman animal species such as chickens to humans, although the US Centers for Disease Control and Prevention (2023) warns "human infection with HPAI A(H5N1) virus have been reported in 23 countries since 1997, resulting in severe pneumonia and death in about 50% of cases." Typically, the infections occur in people who work in close contact with poultry, such as farmers, or people with small flocks of birds.

> **Joe Parish, Medical Anthropologist and Farmer:** When you get into issues of food justice, when you start to take away the power that people have over their own food, you start to introduce practices that can actually make the zoonoses worse. Like the farming of chickens, for instance, which is traditionally done in battery cages and made for very unhealthy chickens, not just for the meat, but also for the eggs, and it just increased the likelihood that you were going to have at some point, and it happened in China, an avian flu outbreak. That's why that occurred, because we were mistreating how those animals were kept.
>
> We were mistreating the animals, and it created the conditions of poor health: depressed immune systems in those animals, and eventually, lo and behold, we get a zoonotic disease. It puts pressure on the animal world, and it's not just the farm animals, although that's a major cause, especially pigs and fowl. Those are the two major ones where we can get very easy transmission crossing the species boundary, but the large firms especially put pressure on the wild animals because they get pushed to the margins ... So we start to put pressure [on animals] by creating situations where the farms are impacting the environment, we're impacting everything in that environment and the animals are the things that get stressed out, it's going to come back to us.
>
> This is likely where COVID came from as well. Putting pressure on those wild bat populations and reducing their habitat, reducing their resources is probably what's causing

> them to react in those ways. It also doesn't help that the population is over 8 billion now at this point, that's also not good, but it's the ways in which we do that, that's problematic. (Interview with Joe Parish 2023)

Global capitalism promotes an extractive economy that necessitates reaching further into nature for land and resources (Singer 2010). Many of the changes that have facilitated climate disruption – the intensification of the agricultural system, the clear-cutting of forests for pastures or to grow feed crops, increased populations in cities – "are accelerating in tandem with the emergence or reemergence of pathogens, climate change, and persistent, large-scale, inegalitarian social structures" (Herring & Swedlund 2010: 7). The spread of infectious diseases is influenced by human activity: people being pushed off the land and into cities, workers having to travel long distances as farm laborers, families needing to travel longer distances to see each other, and so on (Buckee et al. 2021).

On this treadmill of production and consumption, climate disruptions become ever more commonplace and these extreme weather events are more likely to impact the people that had the least to do with behaviors that created the problems. Peasant farmers and people that live off the land are increasingly displaced and are pushed into cities, where, as we learned above, concentrated populations are more "at risk of swift-moving infections and other diseases" (Singer 2010: 32). These can all be seen as urgent economic, political, cultural, and social problems.

In our interview with Vandana Shiva, she discussed rethinking how we use the word "urgent":

> You know, urgency doesn't relate to time. It relates to significance. You know something is urgent means it's very significant ... the relationship of humans as animals to other animals, just layers of constructed hierarchy and constructed separation. And this moment of ours every time the disasters are worse is a time to pause and step back and see what is it that we were made blind to, what was it we were made indifferent to. (Interview with Vandana Shiva 2023)

The United Nations stated that there are some "urgent" issues for decision-makers to consider if they wish to avoid the next pandemic and at the top of the list is "de-risking food systems" (United Nations Environment Programme 2020). They also provide the key drivers of zoonotic diseases, including increased demand for animal protein, unsustainable agricultural intensification, increased use and exploitation of wildlife, unsustainable use of land, changes in food supply, and climate change (United Nations Environment Programme 2020).

Charlotte Blattner, senior researcher and lecturer at the Institute for Public Law, University of Bern, Switzerland, argues that the crux of pandemic preparedness is that pandemics can be prevented, and that we currently are faced with two options: (1) we can fail to change the way we treat animals individually and collectively, essentially accepting that pandemics will always be with us, or (2) we can decide to use the COVID-19 pandemic as an opportunity to "proactively through government regulation, corporate action, and changes in individual behavior" transform the way we treat other animals (2020: 47–8). Similarly, Lymbery argues that "urgent action is needed globally to move away from industrial farming practices that risk public and animal health and welfare in favour of a regenerative, agroecological food system – in harmony with nature – without factory farming and with much less dependence on intensive animal products" (2020: 147).

Michael Greger argues that human beings have "dramatically altered the ecological landscape in which other species and their pathogens must function. Along with human culpability, though, comes hope: If changes in human activity can cause new diseases, then changes in human activity may prevent them in the future" (2007: 278). Given the real health threats posed by intensive production of animals as *food* to nonhuman animals, workers who handle and work closely with animals, and consumer-citizens, it is imperative that we address these health concerns now, before we are faced with even graver consequences.

Building upon these messages of the importance of careful reflection to ensure we initiate the correct course of action is the equally important message from Elder Albert Marshall, who tells us that choosing to do nothing is still a choice (Marshall 2022). Charlotte Blattner (2020) argues that decisions must be democratically decided upon, and as we've discussed earlier, even that would mark a fundamental shift in how decisions are currently being made regarding food and agricultural production. Elder Albert Marshall advises a precautionary approach wherein before acting we determine to the best of our ability how decisions or actions may or may not be in harmony with the earth (Interview with Albert Marshall 2022).

An Incremental Shift to a Compassionate Food Future

Institutionalized settings often have captive populations. Sometimes that is literal, such as when people are incarcerated in prisons. Other times it is more figurative, as people may have few options during a particular time span, such as students in public school or at colleges and universities, or those communities living under food apartheid. Margaret Robinson suggests:

> We need to deal with the food we feed to people who are being held in state institutions. Whether that is incarcerated people who have almost no choice in what they are forced to eat or people who are being held in care facilities or institutions administered by the government, such as children in care. I think when you look at how the state feeds people that really tells you a lot about the people who shaped that [decision]. (Interview with Margaret Robinson 2022)

This issue also relates back to the earlier discussion of food availability, accessibility, and adequacy. Associate professor in the faculty of agriculture at Dalhousie University, Kathleen Kevany suggests that one way to champion a healthier, environmentally friendly, and inclusive diet in an institutional setting such as a university campus, is to make plant-based options the default, so that rather than people having to opt into plant-based options, they would be the default, and *meat* and other *food* products made from nonhuman animals would be the add-on (Kevany 2022).

In the United States, another potential partial solution proposed is the Farm System Reform Act 2019 (I8RA), which is legislation that proposes to move the United States away from large-scale agriculture by banning the opening of new large-scale *farms* and limiting the growth of existing ones in the *meat* and *dairy* sector. This Act can be seen as having short-term goals (banning the opening of new concentrated animal feeding operations [CAFOs]), mid-term goals (limiting the growth of existing CAFOs), and long-term goals (phasing CAFOs out entirely by 2040) (Booker 2020).

Banning intensive farming of nonhuman animals would help create less corporate concentration, and ultimately should lead to more localization in the food system. Such a shift would serve several important functions. First, it would increase the number of farming and food processing jobs, as more farms and local processors would be needed with less corporate concentration and reduced farm sizes. Second, it would help people reconnect with those who are part of the food production. Knowing the farming and food production process would, we believe, facilitate more concern for the workers and nonhuman animals in this system. It is hard to think of a worker as a machine, or a nonhuman animal solely as a product or commodity, when we meet them or know what their lives look like.

Less intensification in the food system has many potential positive outcomes. It decreases the chances of foodborne illnesses or another zoonotic pandemic, and when an outbreak occurs on a small-scale farm, fewer nonhuman animals would be infected and/or would be killed as a result. It would also mean that less *meat* and animal-based products would likely be produced so that this could lessen the amount of animal suffering in the system. We recognize this as an imperfect partial remedy to some of the issues we have discussed thus far. It would be more powerful if this reduction in large-scale animal-based agriculture included policy to help transition farmers in intensive production of nonhuman animals to the growing of crops to feed people, as that could help alleviate some of the issues with food insecurity and reduce emissions from animal-intensive agriculture (Harwatt 2019: 539).

Changing our food production system would not only reduce the possibility of zoonotic diseases, but it would also more broadly benefit our health. It is widely known that diets lower in products made from nonhuman animals, such as legumes, nuts, vegetables, and fruits, tend to be healthier for humans and may contribute to treatment of existing ailments and reduction in the onset of other diseases (Martínez-González et al. 2014; Satija & Hu 2018; Willett et al. 2019; Werner & Osterbur 2022; Betz et al. 2022; 2023).

In contrast, Willett et al. point out that "Western and Westernized diets pose global health risks for humans and the environment" (2019). For consumers, diets that are high in *meat, eggs, dairy,* and several species of *fish*, may contribute to increased risk of heart disease, some cancers, and diabetes (Campbell & Campbell 2016; Satija & Hu 2018; Sterling & Bowen 2019; Szabo et al. 2021). Research suggests that there is now ample evidence that increasing the proportion of diets that are plant-based is beneficial for cardiovascular issues (Satija & Hu 2018).

Many of the scientific articles examined for this chapter define a plant-based diet broadly, and Satija and Hu speculate that "a progressive and gentle approach to vegetarianism ... that incorporates a range of progressively increasing proportions of plant-derived foods" may be the most effective method of persuading people to move in that direction. In other words, a gradual dietary change may be easier for people to adopt, and then to stick with (Satija & Hu 2018).

Werner and Osterbur found that "research suggests that diets lower in animal foods and higher in plant foods are among the healthiest. These diets include vegan, vegetarian, and Mediterranean" options (2022). There is a variety of options within the "plant-based" spectrum of diets, as vegan diets cut out all animal-based products, vegetarian diets cut out *meat* and *fishes* (but allow for *dairy* and *eggs*), and the Mediterranean diet encourages mostly plant-based foods (including vegetables, fruit, nuts, and beans), the use of more healthful fats, and *fishes*, *dairy*, *chickens*, and *eggs* in moderation.

Two common problems faced by advocates of increased consumption of plant-based foods, whether for human health, planetary survival, or consideration of animal protection, is that it is viewed as expensive (and thereby privileged) and as counter-cultural within a Western or Westernized framework. According to Sterling and Bowen (2019), there is a need for more education to counter the common assumption and misconception that it is more expensive to eat a plant-based diet. In their review of several studies that demonstrate affordable plant-focused meals, they give the example that "a plant-based dinner consisting of red beans, brown rice, collard greens, sweet potato, and cornbread could feed a family of four for under $12". But for people raised on a "meat and potato" diet, it can be challenging to picture what cooking and eating looks like if one shifts to an entirely, or even a partially, plant-based diet.

To try and counter those issues, and to recognize people trying to cut back on their intake of *meat* and other products made from nonhuman animals, author Brian Kateman coined the term "reducetarian" and created a foundation, wrote several books, and produced a documentary about its potential to decrease *meat* consumption at a societal level (Kateman 2022: xiii–xiv). His outlook is pragmatic as he views the cultural, political, and economic barriers to people going completely *meat/animal product*-free and tries to counter these with a more flexible approach.

> **Lynn Kavanagh, Farming Campaign Manager at World Animal Protection:** By looking at the issues that people are really concerned about – public health, pandemics, antibiotic resistance, climate change, which are all so negatively impacted by the industrial animal agriculture system or factory farming – we can highlight how we need to make those fundamental or systemic changes. And so, what we're doing is we're really promoting the meat-dairy reduction and more plant-based eating. And in the long term, the hope is that that will result in far fewer animals being farmed and hopefully dismantle the factory farming system.
>
> Because the way I see it, and the way organizationally we see it, is that there's no way to farm animals in a more humane, sustainable way at the current numbers of animals; the earth does not have the capacity to do that, and what's driving the factory farming system is the demand for huge amounts of meat and dairy. So, we need to reduce that ... We'd like to see the government acknowledge animal agriculture as an important source of climate emissions, and as well to set targets, emissions reduction targets, for animal agriculture. Right now, in their climate policy, they do address agriculture, but it's very much more on the technological side, like, let's improve efficiencies on the farm, but not changing what we farm or how we farm. (Interview with Lynn Kavanagh 2022)

In Canada, the Food Guide was updated in 2019 and it now prioritizes plant-based foods. Looking at the guide, one is struck by how far we've come as even the protein portion (which is about a quarter of the plate) shows beans, nuts, tofu, and legumes, along with the more conventional animal-based protein sources (Health Canada 2019).

Lynn Kavanagh, Farming Campaign Manager at World Animal Protection, explained that, within her organization, the Canadian Food Guide is "our low hanging fruit when we're talking to government." Here Kavanagh was discussing the importance of the government promoting the Food Guide as a first step to get Canadians eating more plants and less *meat* and *dairy* (Interview with Lynn Kavanagh 2022). The Canadian Food Guide presents an important educational tool as the website also suggests that people "be mindful of your eating habits," "cook more often," "limit highly processed foods," and recognize that "marketing can influence your food choices" (Health Canada 2019: 49).

We believe that it is important to think about food and sustainability together, as certainly the ability of the land to sustain nutritional needs is connected to the ways in which we use/abuse the land, and the nonhuman animals (both domesticated and those who live independently of us) that live on it. Professor of human ecology, Stephen Wheeler argues that sustainability requires shifting worldviews and these shifts "are transformational, radical, and not yet fully appreciated by most of us that use the term" sustainability (2016).

This means that to have sustainability in food production and consumption, change is required. These changes can be framed as a paradigm shift, as many of our interview participants discussed, so that the planning to create a more sustainable food system requires the following changes: "results-oriented problem solving, a long-term perspective, and holistic or ecological thought" (Wheeler 2016).

It is quite amazing that a huge part of a potential solution to many of the food-related crises we face is literally right in front of us, but at the same time they remain hidden in plain sight. Here we are speaking of looking no further than what is on the menu for tonight's dinner or what is mixed in with someone's favorite coffee. Related to this, Jonathan Safran Foer points out in *Eating Animals* that:

> In the United States, farmed animals represent more than 99 percent of all animals with whom humans directly interact. In terms of our effect on the "animal world" – whether it's the suffering of animals or issues of biodiversity and the interdependence of species that evolution spent millions of years bringing into this livable balance – nothing comes close to having the impact of our dietary choices. Just as nothing we do has the direct potential to cause nearly as much animal suffering as eating meat, no daily choice we make has a greater impact on the environment. (2009: 73–4)

At the same time, when considering the complexity of creating a truly just food system, consumers will not be able to shop their way out of the multitude of issues we currently face (Chomsky 2022). We need a food system based on agroecology, a holistic approach to farming that "seeks to optimize the interactions between plants, animals, humans and the environment while also addressing the need for socially equitable food systems within which people can exercise choice over what they eat and how and where it was produced" (Food and Agriculture Organization of the United Nations n.d.).

Investing in agroecological forms of farming would help to localize farming, employ more farmers, and if governments invested in such small-scale farming the way they currently do with intensive and industrialized methods of food production, could address many of the food justice issues identified in this chapter (Holt-Giménez 2017a; 2017b; FAO, IFAD, UNICEF, WFP & WHO 2018; 2022). For such a transition to become a reality, it would require both political will and grassroots practices and movements to challenge the current industrial system and demonstrate the viability of agroecology (Altieri 2010; Holt-Giménez 2010).

Food as Political: Resistance Is Not Futile

Farmer and author Wendell Berry wrote in his essay *The Pleasures of Eating* that "eating is an agricultural act," and he then goes on to argue that it is also deeply political:

> There is, then, a politics of food that, like any politics, involves our freedom. We still (sometimes) remember that we cannot be free if our minds and voices are controlled by someone else. But we have neglected to understand that we cannot be free if our food and its sources are controlled by someone else. The condition of the passive consumer of food is not a democratic condition. One reason to eat responsibly is to live free. (Berry 2009b)

This is also a theme in the work of food writer Michael Pollan as he echoes Berry's claim by stating that eating "is also an ecological act, and a political act, too" (2006: 11). Meanwhile, Warren Belasco, Professor of American Studies at the University of Maryland, argues in *Food: The key concepts* that, "if you want to create a better future, start by learning to cook. In our quick-and-easy age, it's one of the more subversive things you can do, for when you cook you take control of a piece of the food chain. Moreover, you start to wonder how the food got to your kitchen – and that's a really good question" (2008: 123).

For Indigenous Peoples, access to traditional food is foundational to identity and health, but food sovereignty also directly challenges settler colonialism (Robinson 2019). At the 2019 Vegans of Color Conference, Margaret Robinson explained that "Mi'kmaw values include animal personhood, colonialism intentionally destroys Indigenous food economies, and decolonizing our diet supports food sovereignty" (Robinson 2019). In discussing animal personhood, she explains that nonhuman animals feature prominently in Mi'kmaw storytelling (Robinson 2013; 2017; 2019). She uses Mi'kmaw Elder Wanda Whitebird's words to explain animal personhood: "the people who walk, the people who crawl, the people who swim, and the people who fly" (Elder Wanda Whitebird as quoted in Robinson 2017: 71). Animals are included in Mi'kmaw language as "all my relations" (Robinson 2017: 71).

We have discussed throughout this book some of the ways in which settler colonialism has destroyed Indigenous food economies worldwide. It has also, in some cases strained, and intentionally attempted to break, Indigenous People's relationship with other animals (Robinson 2019). Here Robinson refers to the example of the settlers' hunting buffalo to near extinction in what is now known as the United States and Canada, from a pre-contact estimate of 60 million to a post-contact low of just 1,000 buffalo (Robinson 2019). In relation to the difficulties facing Indigenous Peoples worldwide in trying to maintain or reclaim food sovereignty, they are confined by the expropriation (i.e., stealing) of land for *cattle* ranching and settlements, the forcible removal of Indigenous Peoples from the land, the destruction of ecosystems that supported important traditional food sources, to name but three (Robinson 2019).

Robinson's research has examined the passing down of recipes in Mi'kmaw communities, both traditional ones and those that were blended with settler foods. She explained the importance of this sharing of recipes and food as a form of "anti-colonial resistance in the kitchen where recipes could get passed down because it's a space where you knew the Indian agent isn't maybe watching this closely and people could share food, cooking traditions, even if they could not share their Indigenous language" (Interview with Margaret Robinson 2022).

Decolonizing diets and supporting food sovereignty for Indigenous Peoples globally necessitates that we consider more than individual actions, and examine structural inequalities in income disparity, access to healthy food sources (such as, where are reserves compared to grocery stores and farmers markets), as well as ensuring access to traditional food sources (Robinson 2019). Other research suggests that the "[r]estoration of native ecosystems, revival of traditional food crop cultivation, and revival of traditional knowledge of food preparation, processing, and preservation are important steps to build dietary support strategies against an NCD [noncommunicable chronic diseases] epidemic of contemporary indigenous communities" (Sarkar et al. 2020: 3).

Robin Wall Kimmerer points out that there are thousands of years of science behind traditional knowledge and teachings about Indigenous plants as foods and medicine (Kimmerer 2013). She beautifully captures the reciprocal relationship that can exist between humans and plants when she tells the story of "The Three Sisters" and the significance and interdependence demonstrated in the planting and growing of corn, beans, and squash together, but also how they complement each other when they are eaten together (2013: 128–40). Kimmerer tells us that "They taste good together, and the Three Sisters also form a nutritional triad that can sustain a people" (Kimmerer 2013: 137–8). Complex Indigenous food systems were not 'happenstance', as settlers often incorrectly assumed; they were built upon a longstanding and deep relationship, understanding, and appreciation of the land (Kimmerer 2013).

Transparency and Beyond

In strategizing potential counters to these global food problems that are created by the global economic system, we again argue for the importance of transparency as a necessary part of a healthy food democracy. Adding to this, we consider the importance of education, because knowing what is happening in the current food system is crucial. However, it is not sufficient because, if we want change, we also need action. As Joe Parish explains:

> I think about the consumer, and most people really don't care, they just know what they like to eat, and they want to get that thing. And that's it. That's where their relationship ends with food. I think there needs to be more education about getting people to a point of caring about food. This is especially true when we're talking about eating meat. Most people actively don't want to know how the farming of that animal takes place. Especially the slaughter and processing of the animal for meat. They just don't want to confront that process. (Interview with Joe Parish 2023)

Education on the health benefits of whole foods, small-scale organic foods, and plant-based foods, and how to shop for and cook such foods, would be examples of such education. Since pathways to healthy, sustainable, just, and compassionate food start when we are young, one of the most important places to begin this relationship with healthy whole foods that are culturally appropriate, in our opinion, is in daycares and public schools.

These partial solutions are necessary as we struggle to feed nutritionally dense, culturally appropriate, and humane food to the world's growing human population while facing an unprecedented health crisis with shortages of doctors and nurses, long line-ups at hospital emergency departments, and a reliance on pharmaceuticals to try and fix what ails us. For it to truly make a significant difference in food justice and food inequity, the above-mentioned activities would have to be democratically decided upon, culturally appropriate, and accompany the basic income guarantees we discussed previously.

In our interview with Vandana Shiva, she reflected on what she had learned over the years with her research and community advocacy and said that "I don't see food as a commodity, I see food as the flow of life and the food web as the web of life. Food justice is justice to every being in the web of life" (Interview with Vandana Shiva 2023). As we will see in the next chapter, a shift in both attitudes and actions is urgently needed if we are to successfully challenge the treatment of nonhuman animals in this system, as well as mitigate the interconnected challenges we face as we contend with a global food system run amuck.

5 The Upside Down

The Hidden World of Nonhuman Animals as Food

The cost of a thing is the amount of what I will call life which is required to be exchanged for it, immediately or in the long run.
– Henry David Thoreau

This idea is spreading that species, per se, is morally arbitrary. Yet it still amazes me how deeply entrenched the idea remains in our culture, and so how long it's taking to change things to the way they should be.
– Richard Keshen

Challenging this notion that these animals are economic units and saying that instead they are living feeling creatures, they are fellow earthlings, they are our companions, they are not commodities, they are our friends, not food.
– Gene Baur

October 2 is World Day for Farmed Animals (WDFA). With slogans such as "Stop the slaughter – Go Vegan!," "Non-violence begins at breakfast," and "If you pet a dog, don't eat a pig," this observance offers concerned citizens a way to recognize and grieve the lives lost in the production of *dairy, eggs, fishes,* and *meat*. It began in 1983 and corresponds with Mahatma Gandhi's birthday. Other than this observance, there are very few opportunities for consumer-citizens to critically reflect on their consumptive food choices, nor the implications of the intensive system that produces nonhuman animals for the *meat* industry, their productive output, such as *eggs* and *dairy*, or those fishes factory *farmed* or caught in the sea, that go on to be consumed by humans.

To talk only about human workers, consumers, and the environmental toll of the industrial production of animals as *food* would create an "elephant in the room," or more aptly a cow, pig, chicken, or fish, as nonhuman animals are obviously necessary for this system to exist, but they are often absent in conversations about the system (Adams 1991). Reflecting on the opening quote by Thoreau, "The cost of a thing is the amount of what I will call life which is required to be exchanged for it, immediately or in the long run," we couldn't help but consider this in relation to the lives of so-called *food* animals, that are "thingified" in the creation of animal-based products, *meat* and *fish* (Gibbs 2017). While Thoreau was not talking about other animals in this quote but rather asking the reader to consider the costs of maintaining a consumer lifestyle and whether the amount of work necessary to sustain it was worthwhile, we think it is also relevant when considering the costs of intensive production of animals as *food*. What are the individual and collective costs that result from this commodification?

The Upside Down: Speciesism in Action

For those who haven't watched the hit Netflix show *Stranger Things*, Upside Down is a strange, parallel universe where on the surface things look the same as the regular world, but beneath that surface nightmarish scenarios and monsters are commonplace. There are important parallels to be made with current intensive food production systems because much of what occurs to nonhuman animals, workers, consumer-citizens, and the environment is hidden, distorted, or siloed, and most consumer-citizens have little or only a vague understanding of the externalities it creates.

Without transparency, we have little understanding of what this intensive system looks like, nor do we see the consequences that often result (Singer & Mason 2007; Fitzgerald 2015). Transparency also serves another important purpose as it reminds us that all of the outcomes of this industrialized food production system are connected. Not only does this mean that what happens in one part spills over into the others, but as professor of sociology and critical animal studies scholar/activist David Nibert told us in his interview, we cannot reasonably expect to fix one part of society without being aware of, and working to find solutions for, the issues affecting all parts.

As Nibert explained, "Years ago I began to weave a sociological analysis into what I saw as deeply entangled oppression of humans and other animals. And then the more I learned, the more I realized that the issue of animal oppression can't be disconnected from almost every major issue that's challenging the world today" (Interview with David Nibert 2022). When we went on to ask him how he would demonstrate those common themes to someone unfamiliar with them, he said:

> I guess I'd start by saying that what is happening to other animals is horrific, and there is unimaginable suffering and violence. Most of us have been taught that what is done to other animals doesn't really matter, we are told this is why they exist. We are only told that they should not be treated cruelly without necessity. Few people know we could produce enough healthy and nutritious plant-based food for everyone, and we could begin to heal ourselves and the planet and everyone on it, if our species quickly transitioned to plant-based diets. (Interview with David Nibert 2022)

As he has demonstrated in his extensive research on the intersections of oppression (Nibert 2002; 2013; 2014; 2017) and shared in the interview for this book, in violent conquests and colonialism nonhuman animals have been used as instruments of war to transport soldiers (i.e., horses), as sacrifices to the colonial genocide of Indigenous Peoples in the Americas (i.e., the slaughter of millions of bison on the "US" and "Canadian" prairies), and to occupy stolen lands for *food* production (i.e., when those same lands became held by settlers for profitable ranching) (Interview with David Nibert 2022).

In connecting all this back to his realization that the utilization of nonhuman animals for *food*, for clothing, and for medicine is connected to forms of human inequality, Nibert explained: "I basically try to emphasize the point that you can't deal with racism or sexism or ableism or classism, any of these issues, without realizing that they are connected politically, economically, and ideologically to the oppression of other animals" (Interview with David Nibert 2022). This is essentially framing the treatment of other animals as speciesism.

Speciesism is the elevation of the human species to a privileged position above other animals, and it recognizes and draws attention to the implications this approach has for those animals' lives. Richard Ryder first coined the term speciesism in 1970 to explain the suffering humans cause to nonhuman animals (Ryder 1975: 11–26). He argued that speciesism is a prejudicial attitude that privileges humans over other animals (Ryder 1975). Speciesism in practice recognizes the ways in which the exploitation of other animals is regulated, subsidized, and enforced by social institutions, conventions, and governments. Peter Singer popularized the term in 1975 in his book *Animal Liberation* where he argued it "[i]s a prejudice or attitude of bias in favour of the interests of members of one's own species" (1975: 6).

David Nibert reworked the original definition to include the important recognition that rather than some sort of individual deficiency, speciesism (like all -isms) is "[a]n ideology, a belief system that legitimates and inspires prejudice and discrimination" (2002: 17). This points the way to creating an understanding of this as a systemic form of structural violence that is akin to other systemic inequalities, such as sexism, racism, and ableism (Nibert 2002; Adams 2007; Gruen 2015; Taylor 2017; Harper 2020; Ko & Ko 2020).

In Ryder's analysis he differentiated between the levels of cruelty against nonhuman animals: individual cruelties are the result of ignorance, neglect, or meanness, whereas institutional cruelties are based on "greed, ambition and an unthinking adherence to convention" (1989: 11). While there will be times when an individual worker, transport driver, or handler is cruel towards an animal (e.g., using an electric prod to try and get a downed animal to move) or neglectful (such as failing to provide for an animal's basic needs), it is much more likely that other animals in this system will experience stress or potential harm as a result of the common, and legal, *standard industry practices*. If that is the case, why is it that we know so little about these?

Why We Don't Know What We Don't Know

There is a sense in which people seem to believe that the way we produce nonhuman animals for *food* is natural and inevitable. It's important to remember though that the A-IC is a very recent development in the long history of people often living in relation to other animals. Beginning to peel back the layers and look beneath the surface of common everyday understandings of food and what industrialized production of animals as *food* entails can be an alienating process and can feel akin to culture shock.

Cultural anthropologists explain that culture shock refers to feeling alien in one's surroundings. It is often used to describe the feelings of anxiety or depression we may experience when we visit or move to a new region or country or begin a new role such as changing careers or schools. In researching the relationship between humans and other animals, and especially when examining the literature on what their lives look like under industrialized *food* production, we have been struck by the culture shock we often experience. We do not think that we are alone in this experience, as many of the interviewees we spoke to for this book discussed how nice it was to chat about these issues, especially when discussing their concern for other animals, without being challenged for holding opinions contrary to the mainstream narratives or, worse, being labeled in a derogatory way.

"Carnism" is the term developed by Melanie Joy, professor of psychology and sociology at the University of Massachusetts, to describe the ideology or belief system related to *meat* eating as being ethical, appropriate, and invisible (2010: 30). The invisibility of *meat* eating, Joy argues, occurs as it is presented as not really being a choice at all, and therefore is not open to scrutiny (2010: 10). Relating back to the previous discussion of "-isms," carnism is so widespread and accepted that it is invisible to most people. This invisibility is symbolic, but it is also by design as we typically don't see so-called *farmed* animals' lives or deaths, or those of wild *caught* or *farmed* fishes until they are presented as *food* (Joy 2010).

Joy states that "*We don't see them because we're not supposed to*. As with any violent ideology, the populace must be shielded from direct exposure to the victims of the system, lest they begin questioning the system or their participation in it. This truth speaks for itself: why else would the meat industry go to such lengths to keep its practices invisible?" (2010: 40). In many ways, this invisibility is similar to the US government's ban on photographing the return home of bodies of US soldiers killed in combat. As Grant Penrod pointed out, "The Pentagon's ban on media coverage of the coffins brought to Dover Air Force Base in Delaware looks to many journalists like an effort to control reporting that might bring a negative image to war" (Penrod 2004). The ban was eventually lifted in 2009.

Labeling and legally prohibiting behaviors that are viewed by industries that utilize nonhuman animals as disruptive, such as speaking out against industrial production methods or taking part in undercover investigations of *hatcheries*, factory *farms*, or slaughterhouses, are increasingly being endorsed by governments to protect animal-based industries from public scrutiny (Sorenson 2016). Ag-gag laws, as they are often referred to, are typically state or provincial legislation, originally implemented in some US states, but then adopted in other Western nations, to make it unlawful to disrupt animal industries.

Animal advocates indicate ag-gag laws are a threat to whistle-blowers and to revealing any abuse of animals, or related food safety issues for consumers, in agricultural facilities (Fitzgerald 2015; Sorenson 2016; Gibbs & Harris 2020). Much as the proponents of the so-called Green Revolution have framed such changes in food production as progress, "[f]raming current developments in industrial animal agriculture under this banner of progress has many consequences, but one particular problematic one is that critiquing the industry has come to be viewed as backward or even deviant, and actions are being taken to curtail such critiques" (Fitzgerald 2015: 109).

Camille Labchuk, Executive Director of a Canadian animal law group called Animal Justice, told *Mother Jones* magazine, "What we're seeing now is the backlash from the farm community, based on increased activism," and "I can't say I'm surprised by it because we went down this road in the US. But I'm not sure it's good for them in the end" (Kopecky 2022). What Labchuk is referring to is that in the United States most of the court cases using ag-gag laws have been found to contravene the First Amendment rights and have been deemed null and void. "Animal Justice is now fighting the Ontario [Canada] law in court on similar grounds, namely arguing that it violates Canadians' right to freedom of expression" (Kopecky 2022).

John Sorenson, professor of sociology at Brock University, contends that industries (from the A-IC, chemical and pharmaceutical companies, among others) have a great deal to lose because animal rights and environmental protection groups threaten their financial bottom line, as it "constitutes an ontological, existential threat to those whose identities are constructed through animal exploitation" (2016: 13). Sorenson also notes that this type of discourse makes the current treatment of nonhuman animals appear normal and natural. It also tries to silence counter viewpoints such as those promoting animal rights, and as such it has a great deal in common with other types of conservative and individualistic positions, including "pro-capitalist, patriarchal, [and] racist" ideologies (2016: 13–14).

Such legislation also clearly relates to the ongoing questions of democracy, or lack thereof, that we have raised in previous chapters. Ag-gag laws can be viewed as a threat to having an informed consumer-citizenship and to ensuring transparency within the food system (Joy 2010; Fitzgerald 2015). And, as Joy further notes, "[w]hen animal agribusinesses have become so powerful that they are not only above the law, but also of the law – shaping rather than respecting legislation – we can safely say our democracy has become a meatocracy" (2010: 91).

> **Ceallaigh S. MacCath-Moran, Folklorist and Ethical Vegan:** I was doing participant observation at slaughterhouses briefly right before the pandemic in Toronto. And you should also know going into this, that many years ago I was an over-the-road transport truck driver and so I've driven heavy equipment. I know how to drive heavy equipment; I know what it takes. And so I had very considerable truck awareness around those transport trucks as I was feeding water to the pigs and so forth. And I was concerned at the time, I had read on the Toronto Pig Save website that there were incidences where police were present when this happened too, where what happens at a pig safe in Burlington, or at least it was before Bill 156 became law, was that the activist, there would be an activist in front of the truck with a stopwatch and they would hold up their hands in a peace sign and say, "Please give us 2 minutes with the pigs" and then click the watch and scream 2 minutes, right? And then one minute and 30 seconds, and then clear the truck right, so that was something that they did, and they taught activists how to have truck awareness.
>
> But the truck drivers were, there's a lot of violence ... there's a lot of violence around unlearning common sense [here, Ceallaigh is referring to "common sense" in the Gramscian context]. And we see this in the Black Lives Matter movement. We see this with white supremacists who will not unlearn what we've been fed as white bodied people. And these truck drivers were clearly not having this, you know, because when we bear witness in this way, we draw attention to the suffering of the animals that they're transporting, and it can be seen as a personal affront. And so what was happening, before I arrived in the fall of 2019, was that drivers were pulling forward before the two minutes was up. And if this is the truck and this is the

> activist, they were pushing activists with the truck. And then I was present when this happened ... I have on film, the activist with the stopwatch being pushed by a truck.
>
> It was in this exact spot, four months after I came home, that Regan Russell's body was bisected by a truck when it rounded the corner from the wrong lane and sped into her. And you know, like I was at that site four months before she died. And the truck driver was in the wrong lane, I know this as a driver, was in the wrong lane, swung around the corner too fast and tried to punk her out and she was in her 60s. And I'm 53 and I've got arthritis in my knees, and it can be difficult to get out of the way. And she didn't get out of the way fast enough and he hit her and killed her. And so what I'm seeing here as a researcher is this counter hegemony that's happening in animal rights activism is meeting up against this hegemonic bubble.
>
> And you know ag-gag laws exist to keep that hegemony. And the work of government and animal farmers is to keep this quiet, to keep it silent, to keep it under wraps, to keep people from seeing what's actually happening. And what's happening is that when activists push back, not only are they dealing with trauma, very often they're also dealing with violence. (Interview with Ceallaigh MacCath-Moran 2022)

It is obviously problematic when industries from the A-IC are beyond public scrutiny if we are trying to build transparency into this system, and it is certainly counter to other democratic principles, such as citizen participation and corporate and governmental accountability. Undercover investigations of sites of industrial production of animals as *food* have become increasingly significant in illuminating what is happening in the A-IC; this is important because factory *farms* and slaughterhouses are often sequestered in remote communities, often indoors, and have high fences that would make it challenging for most consumer-citizens to visit or even catch a glimpse of the nonhuman animals in those spaces. Undercover investigations often focus on the most immediate concerns, such as blatant cases of abuse, but they also often offer a glimpse at what *standard industry practices* look like as well (Harris 2017).

Individual Abuses and Standard Industry Practices

The industrial production of nonhuman animals as *food* is a global production system that is so large and intensive that it almost defies comprehension (Harrison 1964; Singer 1975; Schlosser 2005; Pollan 2008; Foer 2009; Fitzgerald 2015; Weis 2013; 2018). Although there is some variation in what is believed to be the actual number, all estimates point to the enormity of these industries. One recent estimate states: "It is startling to realize that humans are estimated to consume roughly 40 million (M) cows, 120 M pigs, 300 M turkeys, 7 billion (B) fish, 9 B chickens, and 64 B shellfish per year" (Saier et al. 2022: 154).

Richard Twine claims that approximately 56 billion land animals are killed for *food* worldwide every year (Twine 2012: 13). Some estimates of the consumption of fishes go even higher and argue that the commercial *fisheries* worldwide killed between a little under one trillion (0.79) and three trillion fishes each year in the period from 1999 to 2007 (Fish Count n.d.). This staggering estimate uses data from the Food and Agriculture Organization of the United Nations (FAO) to convert tons of fishes caught to an individual scale. The FAO reports that 178 million tons of aquatic animals were killed in 2020, including both those in the commercial *fisheries*, as well as those fishes *farmed* in *aquaculture* (Food and Agriculture Organization of the United Nations 2022: 2). These numbers do not factor in the so-called *by-catch* of sea mammals and sea birds, who are inadvertently drowned, crushed, or suffocated in the process of commercial fishing.

While people may be familiar with the enormous numbers of land mammals and birds being raised and killed each year as *food*, many may be unaware of the toll to the oceans in commercial *fisheries* and in *aquaculture*, and fewer still may ever have stopped to consider the process from the perspective of an individual fish caught. Jonathan Balcombe argues that "[f]ishes are, collectively, the most exploited (and overexploited) category of vertebrate animals on earth. Second, the science of fish sentience and cognition has advanced to a point that it may be time for a paradigm shift in how we think about and treat fishes" (Balcombe 2016: 6–7).

Taichi Inoue, Japanese Vegan Activist, Author, and Translator of Critical Animal Studies Books from English to Japanese: One of the issues [is that we have a] transparency problem concerning fisheries because there is [such a] limited number of activists who investigate fisheries. So, it would be hard to know what is going on behind the scenes then, to be able to show people. However, instead, I thought that bringing the agency and the subjectivity of such underwater animals to the forefront is one of the key [ways to] consider the moral standing of these creatures ... as Culum Brown and ethologist Jonathan Balcombe have made a great contribution to understanding the inner world of lots of underwater beings. They give scientific basis for our observation.

Even books or papers about fisheries or fishery science are also helpful. In writing my essay, I used such books and papers to understand tunas' behaviors because I, myself, am not a diver and do not have the opportunity to observe tunas directly. However, instead, there are books about fisheries which describe their escaping behaviors or their resisting behaviors. Of course, such books are a return to capturing and controlling such creatures. However, we can and should appropriate and reclaim such oppressive narratives. And knowing the magnitude of what we deprive them of would be the key to understand the heart part of capturing and killing these creatures. I think that is another method of promoting a sense of transparency. (Interview with Taichi Inoue 2022)

Undercover investigations by animal protection groups of *hatcheries*, factory *farms*, auctions, transport, and slaughterhouses have been a significant tool in raising consumer awareness of inhumane treatment of animals in the industrial production of *food*, and in trying to push for industry-wide welfare changes (Shields et al. 2017). The release of pictures and videos taken by such investigations often makes national news. Some of the recent resulting headlines include: "Shocking farm footage shows piglets with tails cut off and mothers crammed into tiny cages" (Dalton 2022); "Alleged animal abuse in US dairy sector under investigation: Claims of violent treatment and cows being passed off as organic have been presented to the Department of Agriculture" (Kevany 2020a); "Sharks' fins are cut while still alive in sick trade that could wipe species out" (Lubin 2019).

Another example, from the Canadian Broadcasting Corporation (CBC), reported that an undercover investigation found chicks at a *hatchery* in Ontario, Canada, being abused. In "Baby chickens 'cooked alive' at hatchery, animal rights group contends," the undercover video, taken by Mercy for Animals Canada, shows a worker joking about the chicks that had gotten caught in the plastic trays housing them, and were ultimately boiled alive or drowned in the high-powered dishwashers. The video also shows footage of sick and injured chicks being poured into a macerator (a large machine where chicks are ground up alive), and a squeegee being used to push the fully conscious chicks towards the machine's blades (Griffith-Greene 2014).

Interestingly, the comments section of the news outlet garnered nearly 900 reader responses, but those comments demonstrated that while many consumer-citizens may be shocked and outraged when they learn about an individual act of callousness or cruelty towards individual animals, many remain unaware of the systemic treatment of animals which may also cause harm and suffering (Harris 2017). Individual acts of abuse in animal industries may be easy for people to recognize when they see it in a video or news story.

The sample stories in the previous paragraphs highlight some of those actions: hitting or kicking animals who are too sick to stand on their own, downed cows (those that are injured or sick and having trouble standing or walking) being repeatedly shocked with electric prods, drowning chicks in a dishwasher, sharks having their fins cut off while alive and being thrown back into the ocean to drown, and dying cows being piled together and left for hours without medical assistance or euthanasia. But sometimes, the undercover investigations also show institutional practices, known as *standard industry practices*.

In one investigation of a pig *farm*, for example, the investigators documented mother pigs being kept in tiny cages, known in the industry as *gestation* and *farrowing crates*, that are about the size of a refrigerator and where the mothers can do little more than stand up, lie down, and nurse their babies. The same footage showed dead piglets, afterbirth, and tails discarded on the facility floors (Dalton 2022). A common *standard industry practice* is to cut off piglets' tails, euphemistically known as "tail docking," which is deemed necessary when pigs are kept in artificially cramped conditions, such as those in an intensive production facility where they may bite each other's tails due to crowding or boredom.

In reaction to this investigation of two UK pig facilities (that housed both pigs used in breeding and those destined for the *meat* industry), "A spokesperson for the National Pig Association, representing pig farmers [in the UK], said: 'We do not recognise the term "factory farm", and are proud of our animal welfare and environmental standards, especially considering the extremely challenging circumstances pig farmers have been subjected to over the past two years'" (Dalton 2022).

In the story on the abuse at a Canadian *hatchery*, some of the *standard industry practices* seen in the undercover video include newborn chicks being transported on a conveyer belt on an assembly line, sorted chicks being tossed into metal shoots, chicks being dried in a mechanical dryer (rather than under their mothers' wings), and chicks who were injured or otherwise unsuitable for sale being dumped into the macerator and ground up while fully conscious (Griffith-Greene 2014).

Standard industry practices are widely used, and legal. They can include regulations about how long so-called *farmed* animals can be transported without access to food and water, and what methods of slaughter are recommended for various types of animal species and under what conditions. As sociologist and green criminologist Amy Fitzgerald discusses, when the US Humane Methods of Slaughter Act was enacted in 1978, "[t]he industry argued that regulation of meat processing practices was unnecessary and costly. They instead suggested, in line with the escalating neoliberal ethos, leaving regulation to the market: let the consumer vote with their dollar and let competition shape practices in the industry" (2015: 56).

And while there is a law under the Canadian Criminal Code (RSC (1985) c. C46) that protects nonhuman animals from "unnecessary pain, suffering or injury," this law is not typically applied to commercial industries utilizing nonhuman animals for profit (World Animal Protection 2020: 8). Most of the protections for animals used in the production of food in Canada are based on voluntary codes of conduct. The World Animal Protection Organization gives Canada a grade of "D" for "protecting animals used in farming" and argues that "compliance with the National Farm Animal Care Council's Codes of Practice is voluntary, but parts of the farming industry are moving towards second-party verification" (World Animal Protection 2020).

For comparison, Canada fares slightly better than the United States' grade of "E" for "protecting animals used in farming" (World

Animal Protection 2020). The organization gives two examples of some positive change in the Canadian context: (1) several provinces now mention the Codes of Practice in their animal protection legislation, and (2) some animal industries such as the "Dairy Farmers of Canada verifies that all dairy producers comply with the National Code of Practice for the Care and Handling of Dairy Cattle" (World Animal Protection 2020).

One of the reasons for Canada's low grade relates to the voluntary nature of many policies, as, rather than being legislated (and thereby enforced by the state), they are left up to voluntary compliance by the industry's stakeholders. The World Animal Protection Organization also notes that there have been some positive changes written into the Codes of Practice in Canada, such as various updates to animal *welfare*, but unfortunately several important changes to animal *welfare* standards are not set to happen until many years in the future. For instance, the guidelines specify that conditions for egg-laying chickens are to be improved so that they are provided with more space and enriched housing, but the deadline for compliance isn't until July 1, 2036 (World Animal Protection 2020).

Research has pointed to the difficulties of trying to balance all "interests and values of all stakeholders" when it comes to determining animal welfare (Croney & Anthony 2010: E75). In part, one could conclude that sometimes they are simply incompatible, such as the needs of capital versus the needs of nonhuman animals or the environment. Amy Fitzgerald, who has extensively studied the *meat* production system, referred to the regulations that prescribe animal transport in Canada when she said:

> Well, here in Canada, it isn't great, and to be honest with you, it was even worse before they changed the amount of time that you can transport an animal without access to food and water and rest, but we're still behind the EU, [we are] closer in alignment with the US now ... But really, the length of time is quite long, and one thing that I have been struck by in the work that I did when I was looking at documents that were accessed through a Freedom of Information request from the Canadian Food Inspection Agency, [is that] it was clear that the industry was really lobbying hard to have the limits stay the way they were, and not have the number of hours decreased.
>
> And in the end, it seems like the CFIA really, they pushed back to some degree, but in the end gave into their compromise because of pressure from what I think was the poultry industry. The documents

were heavily redacted, of course, but based on the puzzle pieces that I could put together, I'm pretty sure it was the poultry industry. And the animal welfare literature showed that the number of hours should be much lower, that their suffering begins earlier, but the industry was able to lobby to get the number of hours higher than the CFIA's initial recommendation. (Interview with Amy Fitzgerald 2022)

The Canadian Food Inspection Agency (CFIA) livestock transportation document includes a section called the "Regulatory Impact Analysis Statement," which is not a part of the official regulations, but it does provide some important context for the regulations themselves, for the consultation that was undertaken, and in explaining why some of the transport guidelines were amended after industry consultation. The following statement specifically discusses the revisions made after consultation with poultry industry representatives and supports the analysis provided by Amy Fitzgerald. According to the CFIA:

> Some members of the poultry industry have expressed concerns regarding the reduced transport times in the amendments, indicating that it is impractical to provide feed and water to the birds while in transport. They have indicated that in cases where shipments will exceed the maximum times without access to feed, safe water, and rest, the industry will not be able to ship these birds. According to respondents, this will have direct impacts on the profitability of those processors, who could lose the supply of a significant amount of an economical source of lean protein, and indirect impacts on the profitability of producers, who will have to pay to have the birds humanely killed and either composted on site or have the carcasses transported to be rendered. The CFIA met with poultry industry representatives to discuss compliance options for reducing the economic impact of the maximum times. In response to stakeholder concerns, and taking into consideration available scientific evidence, the maximum interval for some poultry sectors was revised, for prepublication in the *Canada Gazette*, Part I, to 24 hours from the originally proposed 12 hours. (Health of Animals Act 2019)

The CFIA seems to be transparent here when it indicates that the original recommendations for transport times for chickens were contentious, and that decisions are not simply about what is in the best interest of animals' *welfare*, when they state: "Most stakeholders agree that regulatory amendments are needed and support the need for them. Opinions, however, are polarized. For example, with respect to the changes to feed, safe water, and rest periods, animal welfare groups believe that the proposed maximum periods without access to feed and water are too long, and the rest periods too short, which would in turn impact the animal's well-being." But, on the other hand, the CFIA argues that some of the representatives from the industries affected by the proposed changes believe that the regulations should be longer, as such changes will have implications for their businesses' profits (Health of Animals Act 2019).

The revised Canadian guidelines for allowable hours of transport, without food and water, are significantly higher than those from the European Union, as is discussed in the following paragraph. And in the report card by the World Animal Protection Organization, it states, "Given that this is one of the most stressful parts of a farmed animal's life, the [Canadian] government should continually update the regulations based on the best available animal welfare science. It is disappointing that the federal Government watered down their initial proposal for decreasing food, water and rest intervals under pressure from industry lobby groups rather than basing them on the latest animal welfare science" (World Animal Protection 2020).

The Canadian guidelines for *broiler chickens* (chickens that are bred for *meat*) and *spent laying hens* (chickens used in *egg* production who are sent to slaughter once their production declines, and are typically used in processed *foods*), for example, allow for them to be transported up to 24 hours without water, and 28 hours without food. In the European Union, "suitable food and water shall be available in adequate quantities, save in the case of a journey lasting less time than: (a) 12 hours disregarding loading and unloading time; or (b) 24 hours for chicks of all species, provided that it was completed within 72 hours after hatching" (Council Regulation (EC) No 1/2005 of 22 December 2004 on the protection of animals during transport and related operations and amending Directives 64/432/EEC and 93/119/EC and Regulation (EC) No 1255/97). In Canada, newly hatched chicks can travel without access to food or water for up to 72 hours from time of hatching (Health of Animals Act 2019).

Whereas certain incidents of cruelty may be dependent on who the individual workers are, *standard industry practices,* like other forms of structural violence, occur regardless of which human actors are involved. These are systemic practices and issues rather than those based on individual people's mistakes, prejudices, or malice.

One example of a *standard industry practice* relates to the *dairy* industry. Because the cow's *milk* is the productive output, cows are repeatedly impregnated in order to facilitate an ongoing *milk* supply.

In fact, three months after giving birth, cows in the dairy industry are impregnated again (Masson 2004: 139). In practice, it means that mother cows give birth, produce *milk*, but do not get to forge a mother–baby relationship, as their *milk* is for the consumer market and "most dairy cattle production discourages all aspects of maternal behavior with the exception of milk production" (von Keyserlingk & Weary 2007: 106). This means that in most commercial *dairy* farm operations, the only maternal (that is, motherly) outcome allowed after giving birth is lactation (von Keyserlingk & Weary 2007).

A study from Quebec, Canada, revealed that newborn calves were taken from their mothers within 12 hours of birth in just over 73% of the farms surveyed, whereas in 32.5% of the farms studied, the calves were taken from their mothers within 2 hours of birth (Vasseur et al. 2010). The study reported that the practice of separating calves from their mothers immediately after birth was even more common in the US, and the USDA affirms that immediate separation happens in close to 56% of US herds (US Department of Agriculture 2008). Weaning creates welfare issues for both the mother and newborn calf because of the distress associated with this process (von Keyserlingk & Weary 2007: 111). And Di Concetto et al. argue that both mothers and their young may experience distress when they are prohibited from maintaining this relationship (2022: 20–21).

In *The Cow with Ear Tag #1389*, Kathryn Gillespie details a conversation she had with a farmer at a *dairy* farm in western Washington. She asked the farmer: "What happens to the calves when they're born?" and the farmer explains that they remove the calves from their mothers within a day. Gillespie asks why this was done and the interviewed farmer explains that "It's better that way. We need to separate them for the good of the cow and calf. The longer they bond, the harder the separation is. You know, it's kind of sad. Even when we remove the calves so quickly, the cows'll bellow for the calves – like they're looking for them – for a couple of weeks a lot of the time. So yeah, it's just better to get it over with quickly so that they don't get too bonded" (2018: 56).

Gillespie uses the terms *"gendered commodification and sexualized violence* to understand the lives of animals in the [US dairy] industry and the discourses that are employed to reproduce its practices" (Gillespie 2014: 1321). From a justice perspective, Allison Gray, Faculty of Social Science and Humanities, University of Ontario Insitute of Technology, argues that in *milk* production regulations, *dairy* cows are often discussed as though they are animal-machines as many of the regulations related to their health and *welfare* are tied to "economic interests" rather than what is in the best interests of the individual cow (Gray 2016: 225).

There is also a series of health issues that commonly face *dairy* cows, in addition to the stress of repeatedly having their calves removed very shortly after birth. The literature on the health effects of intensive *milk* production on a cow's health indicate that lameness is a serious issue (Thomsen et al. 2023), as is the cow's udder becoming inflamed (a common ailment known as mastitis) (Dufour et al. 2019; Hussein et al. 2022: 60). A third common health issue is milk fever. This is considered a metabolic disease which may occur shortly after birth and is characterized by decreased blood-calcium levels, and "significantly increases a cow's susceptibility to mastitis," as well as several other significant health risks including death (Reinhardt et al. 2011: 122).

Such ongoing health issues can lead to a *dairy* cow being deemed to be at the end of her productive life (Reinhardt et al. 2011: 122). Being considered *spent* refers to their productivity waning, not getting pregnant as easily, or contracting a hard-to-treat, or expensive to treat, illness (such as mastitis, lameness, or milk fever) (Gillespie 2018: 57–9). At the farm Gillespie visited, cows have three to four calves, and are milked for nine or ten months a year. They are kept for five or six years and then they are sent to slaughter (Gillespie 2018: 55). This timeframe is often expedited on larger-scale *farms*. The farmer interviewed claimed that "[I]f you're eating a burger from a fast food joint, you're most definitely eating a dairy cow" (Gillespie 2018: 59).

The health risks mentioned in the previous paragraph put into context a rather strange part of the Canadian transport guidelines reminding transport personnel how to determine if an animal is fit for transport. One example stated that an animal was fit for transport to slaughter if they had "no inside body parts outside" and that this could include a "prolapsed uterus or a severe rectal or severe vaginal prolapse" (Government of Canada, Canada Food Inspection Agency 2022). Thinking through what circumstances could allow for a prolapsed uterus or severe vaginal prolapse, it seems likely that we are talking about nonhuman animals that would have recently given birth and/or had several successive births. The guidelines stipulate that cows cannot be transported if they have given birth in the past 48 hours (Government of Canada, Canada Food Inspection Agency 2022). Again, this asks us to consider what reproduction and nurturing looks like for cows in the *dairy* industry, as Dinesh Wadiwel points out:

> A view I have been building over the last few years is that we need to see the capitalist food system as a "metabolic" relation which ties animals, humans and capital together within an interconnected form of circulation. Our food system reproduces hundreds of billions of land and sea animals every year. This requires one form of animal labour; namely, the reproductive or, as Sophie Lewis describes this work, the "gestational" labour of millions of animals who are forced to give birth continually to animals who enter the food system. This birthed "labour force" of billions of animals are required to spend their life engaged in a metabolic labour to produce their own bodies as products, which will be transformed into food commodities after their lives are extinguished. (Blattner et al. 2021: 243–4)

Related to the *dairy* industry and reproductive labor is the concurrent *veal* industry. Male calves born on *dairy* farms are most likely to be sold to *veal* farms. The USDA document *Veal from Farm to Table* indicates that most "calves that are sold as veal are typically slaughtered between the ages of 16–18 weeks." It also notes that male *dairy* calves are used in the *veal* industry because they will not have any value in relation to milk production. The document also points out that the calves are most often removed from their mothers within three days, and that on the *veal* farms, they are typically housed individually (US Department of Agriculture, Food Safety and Inspection Service 2013).

In an online document for *veal* farmers, it reminds them that "feeding calves is an investment, not just an expense" (Calf Care 2019). The same site goes on to demonstrate the disconnect between what is more convenient for the farmer as opposed to what helps comfort the calves:

> It's no secret that bottle-feeding calves takes more time than bucket-feeding calves. Feeding calves milk or milk replacer from buckets is simple; it is easier to pour milk into a bucket than a bottle, and some find buckets easier to clean than bottles. However, when calves drink milk from buckets rather than sucking milk through a nipple, they are more likely to display abnormal behaviours like nonnutritive sucking (sucking on objects within a pen) or cross-sucking (sucking on other calves). (Calf Care 2019)

Both documents fail to consider the emotional implications for both mothers and calves of being separated. Cows are known to be gregarious; they like to be with others of their kind. They are also described in research by Laurie Winn Carlson as "nature's most protective mothers" (Carlson, in Masson 2004: 137).

As mammals, calves are supposed to be nourished by their mothers. And when given a choice over their own lives, female cows "stay with their mothers their entire lives, while males leave the herd after they reach one year of age" (Berreville 2014: 188). When examining these *standard industry practices*, it is important to note that they are dictated by economic decisions, rather than animal well-being. And in their economic assessment of *farm* animal *welfare*, Norwood and Lusk remind consumers that regardless of whether they consume *veal*, they are supporting (and "subsidizing") its production when they consume *dairy* (2011: 144).

Animals used in the production of food are often constructed as unemotional, and framed as though they don't mind the way that they are being treated. Marc Bekoff, Professor Emeritus at the University of Colorado and a behavioral ecologist and cognitive ethologist, explained that it is not that animals such as cows are unemotional, it's just that we as humans don't recognize or look for those emotions in *farmed* animals. In our interview he explains:

> I think all the work that's being done in cognitive ethology, on animal emotions and sentience, is beginning to make inroads. Cows aren't dumb, but to decrease cognitive dissonance, people create false characterizations of these intelligent and sentient animals. People will say, well, cows aren't very emotional. I say they're very emotional. First of all, they have very expressive faces, but we don't know much about them because we focus on dogs and cats. And moms and their children don't like being ripped away from one another. Of course, we know this from Harry Harlow's horrific experiments when he was raising monkeys on wire mothers, and they went crazy. They're mammals. They have the same nervous systems, the same neurotransmitters, and the same emotions as we have. It's just we're not familiar with them. (Interview with Marc Bekoff 2023)

Interestingly, people often try to have it both ways. They claim that nonhuman animals don't really mind being housed in cramped, boring, isolated places but, on the other hand, they may be thinking about or accepting the need for changes that improve *farmed* animals' *welfare*. Bekoff reflected on this contradiction: "If people really didn't think these animals suffered then there wouldn't be moves to modify and tighten up animal welfare regulations" (Interview with Marc Bekoff 2023).

Gillespie discusses the "doublethink" necessary to shield ourselves from the reality of what happens to other animals in the production of *food* (2018: 145–9). The phrase is borrowed from George Orwell's novel *1984*, and Gillespie argues that consumer-citizens are exemplary at holding opposing viewpoints on animals at the same time and accepting that both are true, and that it should feel familiar because it is commonly employed in people's day-to-day lives (Gillespie 2011: 121).

Every time we engage in ignoring or denying something, we engage in doublethink (Gillespie 2018: 146). For instance, some consumer-citizens are uncomfortable eating other animals after they learn about the industrial system and the lives of nonhuman animals within it. They may seek out alternatives, such as *humane meat*, and even though they may know that many of the same *standard industry practices* exist in that productive system, and that the nonhuman animal will still be slaughtered to become *meat*, they can ignore that reality and feel good about their decision to purchase *humane* animal products (Gillespie 2018: 146–7).

It reminds us of the term "ambivalence" used by sociologists Arnold Arluke and Clinton Sanders in their 1996 book *Regarding Animals*, where they build on the work of sociologist C. Wright Mills. In it they discuss the myriad ways that people are able to balance opposing viewpoints of nonhuman animals simultaneously and remarkedly not see this as a contradiction at all (1996: 4–5). They argue that, from a sociological perspective, inconsistent viewpoints and treatment of other animals are the product of social forces, such as hierarchical ways of classifying others (Arluke & Sanders 1996: 167–8).

In Gillespie's work she speculates, and hopes, that telling the stories of individual animals – like a mother having her calf taken away only a few hours after birth, or a chick dies in the dishwasher shortly after hatching, or from her research when she relays watching a young calf at a *livestock* auction trying to nuzzle an auction house worker in the auction ring and watching in horror as they are hit in the face with a paddle and told, "I'm not your mother!" (2018: 196) – may help consumer-citizens to think through how we currently treat other animals and how we could significantly challenge this and create something radically different (Gillespie 2018: 219).

While not a radical shift in people's relationship with other animals, there has certainly been a great deal of interest in alternative farming methods that are utilized to try and reframe the production of other animals for *food* as *humane*. Do such production methods provide a guilt-free way to continue eating animal-based products? What do such industries look like for the nonhuman animals involved? Do such methods offer a solution to the problems in the A-IC outlined in previous chapters? It is with these questions in mind that we now answer the question, is there such a thing as *happy meat*?

Happy Meat?

What is framed as *humane* production of nonhuman animals in the industrial production of animals as *food* does not transform the human–animal relationship or eradicate many of the issues outlined in this book. Theoretically there may be the potential for greater animal *welfare* if farming introduced higher standards of care, but there are a great many changes that would need to be made, and many of those changes would simply be incompatible with the profit-seeking model of intensive production. At best, it ensures that animal *welfare* guidelines are being followed, which could be argued should be the case anyway, and at worst, it simply gives consumer-citizens a false sense that the *meat* or animal-based product they are consuming comes from a *happy* animal, charges them a premium for that feeling, and justifies the continuation of the status quo of this hierarchical relationship.

In the journal of *The American Society of International Law*, Saskia Stucki argues that "[m]any humane labels are notoriously vague, unregulated, and unenforced, with no meaningful content or oversight and welfare standards that do not (significantly) go beyond reiterations of the legally required minimum or reflections of standard agricultural practices" (2017: 278). Many animal protection organizations agree (e.g., Animal Justice 2014; Humane Canada n.d; ASPCA n.d.).

The overwhelming majority (99%) of *meat* in the United States comes from conventional slaughter facilities and intensive *farms* (Gillespie 2011: 110); "[t]he other one percent is accounted for by alternative, often small-scale, family-run, meat producers. These farms raise animals labeled as 'free range,' 'grass-fed,' 'organic,' 'natural,' or 'cage-free'" and that "[t]hese terms are all used to advance the discourse of 'happy' animal *lives* (while ignoring animal deaths) and represent a continued resistance to referring to slaughter in labeling" (Gillespie 2011: 110). Nonhuman animals that are raised under potentially higher standards will still in most cases have to be transported to slaughter facilities outside their locales, and transport is known to be a stressful experience for nonhuman animals.

In highlighting what each of these terms means, Gillespie explains that *free range* suggests that animals can come and go from the indoors to the outdoors as they please. The USDA defines *free range* as providing "continuous access to pasture during the growing season" (U.S Department of Agriculture n.d.). Consumers should not assume that this means that the animals have access to wide open spaces, that they can access those spaces at any time, or that they are routinely encouraged to do so (Gillespie 2011: 110). The term *grass-fed* simply means that cows, for instance, are fed "grass and forage" and Gillespie notes that USDA regulations stipulate that they must have "access to the outdoors during growing season," but that in many jurisdictions, that could be less than half the year due to constraints such as weather. *Grass-fed*, in the US context, also does not mean hormone- or antibiotic-free (Gillespie 2011: 111).

The certified organic label means that foods may not be genetically modified and animals under this label must be fed organic feed, have access to the outside, and be cared for without antibiotics or growth hormones (US Department of Agriculture 2016). Gillespie reminds us that access to the outside, as she discussed for *free range* animals, may not look the way we imagine and may not translate into animals having easy and ongoing access to the outdoors (2011: 111). So-called *natural* animal products, such as *natural meat*, refer only to the fact that nothing artificial has been added to the product and that it is only minimally processed (US Department of Agriculture 2023).

Finally, the designation of *cage-free* is most often used to describe the conditions of chickens. The term *cage-free* suggests that the birds are kept in areas where they are able to move around and are not in battery cages. Unfortunately, it does not mean that the space must include access to the outdoors. In Canada, similar food labels exist and, for the most part, they utilize similar definitions (Animal Justice 2014).

Most of the food labels that apply to animals used in the production of *food* in Canada are also unregulated (Animal Justice 2014). Most of the *standard industry practices* we discussed previously, such as chicks having their beaks seared off or chickens that are raised for *meat* being killed at around four to ten weeks of age and, by the end of their lives, living in a density of "about 100 square inches (0.7 square feet) of space" per chicken, are still common (Norwood & Lusk 2011: 129).

The European Institute for Animal Law and Policy has created a comprehensive comparison of *standard industry practices* as compared to organic standards regarding animal care (Di Concetto et al. 2022). They confirm that only the guidelines for the *organic* certification (and not the other types of quality certifications available in Europe) include rules specifically related to animal *welfare* (Di Concetto et al. 2022: 4). It is also important to note that the EU is already ahead of both Canada and the United States in its *welfare* regulations for *farmed* animals, as "[a]nimal welfare in the European Union is measured relatively holistically, and includes health, productivity, physiology, and ethology indicators. The United States, conversely, officially uses only health indicators" (Fitzgerald 2015: 91). But even with the considerably more holistic view of *welfare* officially taken in the EU, some of the "standard industry practices" related to a lack of maternal care in *dairy* and *egg* farms are still allowed under the *organic* label (Di Concetto et al. 2022: 21).

Sophie Riley, Associate Professor, Faculty of Law, University of Technology, Sydney, Australia: There is this idea that you treat animals as a whole. So, in other words, if you're looking at egg production or sheep or cattle, even though they're made up of individual animals, they're treated as a bulk commodity. So, the first thing for me would be to have these animals treated as individuals. Because sentience is felt at an individual level, not at this bulk commodity level. So, what would that mean? That would mean that food products would become more expensive to produce. And that would have to be passed on to the consumer, so there'd have to be some sort of way of easing that in, so it didn't affect the people that are making a living out of it and the people that are relying on the animal products to eat.

For example, it would require an awful lot of money and I don't know whether governments would be prepared to do

> that. But it would mean bringing into the open all these hidden costs to the animal. And hidden costs of the industry such as externalities for the environment. It would mean having a good conversation and then looking at the ways in which these problems can be remedied. You know, don't try and fix everything at once because you're not going to be able to do that and you probably won't know what needs to be fixed until you start getting into it and then deal with one thing at a time.
> It could take decades, but you've got to start somewhere. There's a lot of political problems with that, though. Huge political problems. And I just wonder whether people themselves, whether society can give that a bit of a push. And then if you get the industry on side? That's half the battle, no matter what the government decides. (Interview with Sophie Riley 2023)

The meaning behind the labeling of products made from other animals is not easy for consumer-citizens to navigate. As we saw above, there may not be common understandings of words or behaviors. Marc Bekoff uses the term "humane washing" (developed with Jessica Pierce) to describe the corporate use of the term *humane* animal products (2016). Like the term "greenwashing," it describes the process of companies trying to appeal to consumer sentiment by convincing us that we can positively affect animal *welfare* or environmental sustainability by buying products (Bell & Ashwood 2016: 65–7).

Both greenwashing and humane washing rest largely on the premise that people are first and foremost consumers, and the commonly held belief that the only way for citizens to influence what happens in institutions and economic systems is by exercising their right to vote … with their spending dollars (Chomsky 2022: 75–7). Bekoff argues, "if you hear the word 'humane,' you can pretty well bet that something bad is happening to animals and somebody is trying to clean it up and make it look less ugly as 'humane' is a dirty little lie" (Bekoff 2016).

While farms claiming higher *welfare* may be happy to discuss their *humane* treatment of animals under these largely voluntary standards, discussions of the animal's death are less likely to come to the forefront. Because when consumers are confronted with the realities of slaughter, even when purported to be *humane*, they may be less likely to want to consume *meat* (Gillespie 2011: 116). Marc Bekoff provides an example of humane washing:

> Welfare practices are essentially arguing that these animals are treated the best we can "humanely" treat them so then, it's okay to use them, which is actually the reason why Jessica Pierce and I wrote *The Animals' Agenda*, where we developed the concept of animal well-being which focuses on *individual* animals. This view focuses on each individual's pain and suffering, and you can't humane-wash or whitewash it by saying, well, we treated these cows well. It's like Temple Grandin saying I'm doing the best I can, so it's okay to torture these animals. No, it's not, Temple. (Interview with Marc Bekoff 2023)

Here Bekoff is referring to the modifications to some industrial slaughterhouses developed largely by animal behaviorist Temple Grandin, who is considered an expert in *farm animal welfare* and *humane* slaughter methods. These modifications are portrayed as a positive development in industry and mainstream narratives of animal *welfare*.

Some of Grandin's designs include a curved path to the kill floor (so that animals don't see what is about to happen to them until the last moment) and the upright restraint that holds cows still while they are stunned by a captive bolt to the head in a regular slaughterhouse, or that holds them in place while their neck is tilted and their throat slit while conscious, as is the case in jurisdictions that still allow this form of ritualistic slaughter (Bekoff 2016; Grandin 2018).

In discussing the restraint system, Grandin frames it as "[a] very humane position for cattle. Cattle are restrained in a comfortable, upright position" (Grandin 2018). Of course, Grandin is right if you are comparing the *welfare* of these animals and the two choices are: be hoisted up by their hind legs and having their throat slit while thrashing around in the air; or being held in place and having their head tilted up and then having their neck slit. In one of the essays on these restraints, Grandin includes illustrations and pictures of the machines, nonhuman animals in them, and animals after they are killed. One picture of a cow in the upright restraint, presumably in the moment before their death, shows an animal whose eyes are wide, in what we interpret as a look of complete terror.

In her book *The Ultimate Betrayal: Is there happy meat?*, Hope Bohanec defines *humane* "as being characterized by tenderness, compassion, and sympathy for people and animals, especially for the suffering or distressed. The definition is completely antithetical to the act of treating animals as commodities instead of as sensitive and emotionally complex individuals" (2013: 6). Thinking back to the voluntary regulations and the cost–benefit economic equations being applied to nonhuman animals' lives and suffering, this definition is fundamentally different from how the term is currently used in *animal-based industries*.

In their book, *Compassion by the Pound: The economics of farm animal welfare*, F. Baily Norwood and Jayson Lusk examine "farm production practices" and "farmed animal welfare" through the lens of economics, essentially giving the reader a cost–benefit analysis of *welfare* for animals used in the production of *food*. Near the end of the book, they tell the reader, "If there is one salient fact we have learned talking with thousands of people about farm animal welfare, it is this: *people do not know much about the way farm animals are raised*" (2011: 327). They later argue that "ultimately it is *consumers*, not farmers, who decide how farm animals are raised" (Norwood & Lusk 2011: 355).

These two viewpoints don't seem to coincide very well, nor do they seem able to facilitate the kinds of change needed to build a compassionate and democratic food system. As we learned earlier in this chapter, it is not by chance but rather by design that most of what consumers are taught about animals' lives in the production of *food* are only partial truths, and that it is increasingly difficult to learn what is really going on, and what is known is often dictated by industry ads and spokespersons (Hannan 2020).

Research does suggest that access to multiple forms of information on industrial production methods can shift people's perceptions of animal *welfare*, especially reinforcing the beliefs of people who already hold pro-animal ideals. Ryan et al. found that participants were less willing to accept *gestation stalls* for pregnant and nursing pigs after viewing multiple and varied forms of information (such as print material, videos, etc.) on *sow* housing. For example, prior to receiving this additional information, 30.4 percent of respondents supported the use of *gestation* stalls, but this number dropped to just under 18 percent after they had a chance to view the information provided. The more educated consumer-citizens become on issues of nonhuman animal *welfare* within *food* industries, the more resistance we should expect to such *standard industry practices* (Ryan et al. 2015: 1).

Perhaps also, as retired anthropology professor, science writer, and public speaker Barbara King points out, as people learn more about the other animals constructed as *food* under intensive agriculture, such as their unique capabilities, emotional complexity, and intelligence, they may act differently in the world (King 2021: 175). It is with this in mind that we look at a more joyful and compassionate appreciation of animals constructed as *farm animals*.

Turning the World Right-Side Up: Recognizing Nonhuman Animals for Who They Are

While it is obviously important for us to have a clear understanding of what the variety of species of nonhuman animals' lives look like under the A-IC, in addition to such transparency, it is also crucial for us to see them for who they are: individuals who are living, breathing, feeling beings that take part in emotionally rich relationships and have complex inner lives (Masson 2004; 2009; Balcombe 2006; 2010; 2016; Bekoff 2007). We did not want to end this chapter leaving the impression that animals' victimization is all that we need to understand. Because, as Jonathan Balcombe advises, it is essential for us to remember that animals "are not just living things; they are *beings with lives*" (Balcombe 2010: 203).

Balcombe suggests that we take a bit of time to reflect on this:

> The next time you are outside, I encourage you to notice the first bird you see. Make a mental note that you are beholding a unique individual with personality traits, an emotional profile, and a library of knowledge built on experience. You may be looking at a majestic bald eagle or an ordinary house sparrow. It makes no difference – what you are witnessing is not just biology, but a biography. (Balcombe 2010: 203–4)

Part of this really seeing nonhuman animals involves appreciating the parts of them that may be similar to each other and to humans, but perhaps more importantly is to feel that same appreciation for the differences. In scholar, artist, and disability rights and animal rights activist Sunaura Taylor's book on animal and disability liberation, she examines how caring for nonhuman animals should not be dependent on them being the same as humans, any more than that should be the case between human groups and individuals (Taylor 2017). Attorney, journalist and author Jim Mason writes in *An Unnatural Order* of the importance of the stories we tell about other animals and how often we compare them to us, as humans, and how in modern times this often means that they come up short in those estimations (Mason 1998). He argues that "[t]he West's agrarian culture, more than any other, subjugated animals and exalted humankind" (Mason 1998: 98).

We tend to think of humans as part of a culture and society, and other animals as purely instinctual. But it is much more complicated than that as we all, human and nonhuman animals, have primary emotions that are evolutionary, alongside secondary emotions that are more complex and take into consideration situational contexts in which we find ourselves (Bekoff 2007: 7–9). Bekoff cautions that "[e]motions are the gifts of our ancestors. We have them and so do other animals. We must never forget this" (Bekoff 2007: 166).

Comparing nonhuman animals to us is patronizing and fails to recognize the important scientific fact that "[a] differently intelligent animal may experience life just as intensely as does a human" (Balcombe 2010: 19). Even when there are evolutionary traits within species, we all exhibit individual characteristics and personalities that, while influenced by genetics, are variable and do change over time because of environment, experiences, and relationships: "there's no the 'dog', the 'coyote', the 'human.' You've got to concentrate on individuals. And part of the egalitarian view would be to respect the individuality within each and every species" (Interview with Marc Bekoff 2023).

When we consider other animals only as a group or species, we miss so much about them as individual beings – they are emotional, perceptive, sensitive, and intelligent – and for all these reasons and more they deserve "our deepest concern and consideration" (Balcombe 2010: 78). And perhaps most amazing, as Masson points out, is the fact that "Farm animals – in spite of or perhaps because of the fate that invariably awaits them – seem able to retain their capacity for deep feeling, including, miraculously, their love for us" (Masson: 2004: ix).

There are so many stories that demonstrate the rich lives of so-called *farmed* animals but due to space constraints we will share just a couple here. As Balcombe explains, "Cattle, when first let out into the fields following a long winter confinement, tear about the field, kicking their legs into the air. They seem literally to be full of the joy of spring, and look for all the world like excited toddlers released into the playground after hours sat behind their desks" (Balcombe

2006: 166). Similarly, Marc Bekoff, who has studied extensively the play in nonhuman animals, provides the example of pigs using "play markers such as bouncy running and head twisting to communicate their intentions to play" with one another (Bekoff 2007: 97).

The final story comes from Jeffrey Masson's book *The Pig Who Sang to the Moon: The emotional world of farm animals.* In this story he describes talking to Marilyn Waring, a former member of the New Zealand Parliament and professor at Massey University in Auckland, about her relationship with her companion goats. In it she describes them as fun-loving, loyal, cheeky, and wise, while also highlighting the similarities and differences between the goats (Masson 2004: 124–8).

In one story she tells Masson about watching her goats playing and says that it "must be seen to be believed. I remember once watching them running downhill to a corrugated roof which was followed by a few meters of bare ground, then another roof. How they loved to hear the sounds of their hooves on the roof. The thing is, they would actually take turns. They did not all come at once, just one at a time. The others would wait in line" (Masson 2004: 126). This corresponds with research that suggests "Animals love to play because play is fun, and fun is its own very powerful reward" (Bekoff 2007: 94).

As Bekoff tells us, nonhuman animals do not typically seek out activities that they do not like, and the desire to play, and "[t]he joy associated with play is so strong that it outweighs the possible risks, such as injury, depletion of energy and therefore compromised growth, and death by a perceptive predator" (Bekoff 2007: 95). It does not take a huge stretch of the imagination to understand that these moments are the ones that make life worthwhile, and without them nonhuman animals are not able to experience important forms of well-being.

There is a more-than-human world that is all around us and it's important to remember that even though we tend to think about other animals as species, they are individuals who experience the world as such (Balcombe 2006: 210). We just need to recognize and value those lives. The reality of the industrial production of animals as *food* does not look the way that we have been socialized to see it (red barns, rolling green fields, and animals roaming about with others of their species).

This system, and the individual animals that reside within it, are hidden, and often deliberately. The implications of this system do not stay put in one locale, as they are externalized into other areas. In this chapter we have focused on what the spillover looks like for individual animals. We have also highlighted the reasons why we should care about the lack of transparency and we have examined the importance of knowing how *food* was produced, and why it matters in creating a more democratic, just, and compassionate food system.

While the raising, and killing, of other animals for *food* is nothing new, the intensity of such industries today is unprecedented (Cudworth 2015: 14), and, as we've seen throughout this book, so are the consequences for humans, nonhuman animals, and the planet. As we move into the final chapter, we focus on some of the positive changes that have already been implemented in the food system, and those that could be created in championing a truly democratic and compassionate food system.

6 Towards a Compassionate Food System

We are all part and parcel of this wonderful creation. Everything we do must be done in harmony with nature. We have to sustain ourselves but we shouldn't compromise the ecological integrity of the earth. We must maintain and advance biodiversity. No creature is of less value than a human. Nature doesn't need us but it does need other animals. Let's use our intelligence and knowledge to maintain the earth for others.
– Mi'kmaw Elder Albert Marshall

We began this book by asking the question of how we can better promote and work towards democratic and compassionate systems of decision-making and food production globally. In order to address this question, we have used a lens of climate change to examine the structural violence of the industrial and corporate food systems and how these intersect with various other forms of oppression and marginalization. When we look at these issues, we are reminded of Walt Kelly's infamous cartoon released on Earth Day in 1970, a satire drawn from the US War of 1812: "We have met the enemy and he is Us." Consequently, given this reality, it is perfectly understandable that we might experience denial and grief as we realize there is no single clearly identifiable cavalry coming to save the day.

Perhaps today we could respond to Kelly's sentiment with the expression: "We have met the rescue team and it is Us." While this idea may lead us to panic and run screaming from the room, it is actually good news. Any one of us, anywhere, can make a difference – on the spot *today* – wherever we find ourselves; whether as a mother lovingly breastfeeding their baby or as a spokesperson on the floor of the UN General Assembly. We are all part of various shapes of family, community, workplaces, institutions, and movements engaged in their own ways of trying to live a better life.

We have emphasized the need for processes and institutions that pre-emptively consider effects on communities, nonhuman animals, and nature as a "precautionary principle," *prior* to engaging in any efforts towards "development" or economic prosperity. We have also argued throughout this book that the two pillars – just food systems and democratic accountability – are inseparable. Localized, community-based decision-making at the site of production, any production, helps communities have increased control over what is taking place on their land and in their communities.

In addition, a shift away from having to depend on expensive, fossil fuel-burning production and transport systems, which are now untenable from a climate change perspective, is also needed. While it is true that many cannot produce all the food needed in their own communities (soil quality, climate, geographic location, etc.), there are ways to move in concentric circles from the local outwards, always deferring to the local where possible. A movement towards always choosing what is closest and most harmonious in terms of relationships to the land, workers, and other animals is an important start.

There are new ways of thinking and acting, many drawn from ancient traditions and historical struggles, that can help us build just societies; that could move us towards living on this beautiful planet in healthful and ecologically respectful ways.

Radical Democracy?

A key foundation to these ways of being is more grassroots, localized forms of democracy that ensure democratic accountability on the land where we find ourselves most immediately. And it is important to repeat that the kind of democracy we are referring to is not ultimately attached to Western liberal electoral systems, the ones that are so greatly failing us today. We can decolonize the whole idea of democracy by recognizing that it can involve an expansive and diverse set of values and practices rooted in the histories, cultures, and traditions of peoples around the world.

This idea should not be considered "radical," because it simply recognizes that cultures around the world have for millennia found different ways of expressing their desire to be free while living healthfully and safely in community. However, in contemporary capitalist democracies, it is indeed a radical idea. The etymology of the word radical is derived from the Latin "ladic" meaning "root." According to the Merriam-Webster dictionary, it means "relating to or proceeding from a root."

We could take this as a metaphor for a very simple concept. Firstly, that democracy has roots in cultures around the world and any new forms should be expressed in culturally specific and appropriate ways. Secondly, that decision-making is, or should be, rooted in the individual people, in the specific communities on the specific land with the other specific species affected by the decisions being made in a particular place. And, thirdly, this expansive idea of democracy, as Indigenous friends remind us, always takes into consideration the foundational relationship to the land, recognizing that decisions made in specific places affect others sometimes living in lands far away.

This third pillar requires something beyond the important task of taking care of our immediate families and communities. As Mi'kmaw Elder Marshall often says, and as the reality of climate change loudly reminds us: nature has no jurisdictions (Ongoing dialogues with Elder Marshall). This approach encapsulates our sense of what a more *radical* idea of democracy entails, requiring that circles of care and compassion extend beyond geographically or jurisdictionally defined regions.

As we have seen, climate change shifts the perspective from "this is something that would be great but is unrealistic" to making it a global imperative. However, while it is a simple idea, this spirit of a community that extends to "all my relations" and to mother nature herself, as we have seen, is countercultural, radical, and perhaps even revolutionary in the context of twenty-first-century global capitalism.

The spirit of democracy speaks to *meaningful control over the decisions which affect our lives*. Ideally, it would also ensure, to the best of our ability, a system of protecting the lives of all species and a harmonious relationship with the land. Our guiding assumption is that in promoting this vision of radical democracy we can, and perhaps should, go first in protest and advocacy to the areas where we see the worst harms being done, such as the myriad violations against the rights of humans, nonhuman animals, and the violations to natural systems identified throughout this book.

The first assumption should be that we act with the precautionary principle in mind by attempting to do the least harm possible every step of the way. In embracing the precautionary principle in all future activities, we become committed to doing the least harm. Intersectionality and cross-movement solidarity imply that as educators, students, activists, economists, politicians, etc., we can focus on specific areas based on place, knowledge, experience, and identities.

While it is beyond the scope of this chapter, and indeed this book, to provide a comprehensive survey of alternatives emerging globally, we hope to provide a sketch of some of the key principles and practical approaches that may guide us in thinking, being, and acting in the creation of a just world. In this analysis, as always, we emphasize that food justice is just one part of a much broader struggle for social change.

How Do We Initiate Positive Change? From Paradigm Shifts to Building Community

With an eye to suggesting potential solutions to the myriad crises identified, we want to reflect on what we have learned from the participants in this project. Their voices have assisted us in understanding and framing the broad areas connected to building justice and democratic principles in the production and consumption of food. In reflecting on some of the key stories and ideas from the interviews, we have identified three emergent themes: *paradigm shift, community building,* and *initiating effective change*. Rather than being distinct categories, there is an important overlap, and the interviews provided a great deal of food for thought, so to speak, as we consider ways to progress, as well as reminding us of what is at stake.

Paradigm Shift

While participants did not all advocate for the same "paradigm shift" or even always use that term, looking deeply at their transcripts and comparing them allowed us to find multiple points of intersection and interconnectedness between them. In thinking through what a paradigm shift might look like, it is important to begin by first acknowledging that it will necessitate a change in perspective from the broadly accepted understandings/assumptions about the way the economic and political systems are currently organized.

So, rather than an extractive system built on the instrumental premise that we must have continuous growth, with the accompanying commodification of land, nonhuman animals, and humans that this entails, the shift will by necessity involve a longer-term orientation. It will also involve consideration of the important Indigenous principle of Seven Generations, reminding us to continuously consider the implications of both individual and collective actions on those born seven generations ahead.

Participants, such as the first Mi'kmaq Member of Canadian Parliament Jaime Battiste, discussed the importance of this long-term framing in decision-making while also recognizing the difficulty that exists in getting people to acknowledge its importance. What Battiste's stories reflect is the fact that within the context of contemporary political systems, and the actually existing way in which "democracy" takes place in many countries today, it can be especially difficult to consider the long term and to plan for it in connection to deep structural inequalities and climate change.

Not only are you as a representative often in a short-lived term in office guided by specific party-identified priorities, but you are also facing the fact that most citizens are worried about their current and most immediate realities, such as financial instability or crisis and, of course, really basic but daily things like potholes!

> **Jaime Battiste, Member of Parliament, Sydney-Victoria:** The question that frightens me the most is: are we doing enough? Have we done enough? And, if not, how can we make people realize that we are in the midst of financial instability and a global crisis, everything from pandemics to war? How can we make them see the real problems lying ahead of us? And saying that we need to value this, and we have to make sure that we're doing what we need on this and that our response is in time. I ask people to wake up together and say that we need to figure out how to do this together. But, you know, it almost seems that within many spheres, people are only seeing what is directly in from of them, it is a very short-sighted view, this year or next year, when the looming crisis might be eight years away.
>
> But places like up north in Canada that are starting to feel the effects of this. We did a study on food security in the north as part of the Indigenous Northern Affairs Committee and all the things that we are fearing about what climate change is going to mean to us in the future is already happening there with the

> loss of food security and with the loss of ice bridges they use to carry food to folks. It is not something in the future there, it's happening now and it's requiring our government to take increased steps to say, "Here's what's going to happen."
>
> And if anyone who wants to see, you know, we're beginning to see devastation like the fires in British Columbia and the floods that we've seen in Cape Breton, and everything that's happening in the North. These are *already* [participant emphasis] happening in our coastal communities, and we have to wake up to realize what is happening, and it is often difficult because people don't want to look past their personal comfort zones. That's the challenge that we have as humanity: how do we get past our own temporary comfort zones to realize what's ahead of us and the possibility that by the time we choose to act, it might be too late? (Interview with Jaime Battiste 2022)

Regarding a paradigm shift, the importance of humility – when looking at the broad issues we face in building a just food system and a just society – was also often mentioned. For some participants this involved a reflexive positioning of themselves in their work by stepping back and asking, "Are we doing enough?" And "Are we being effective?" Underlying many of the participants' responses, some key prerequisites to a just food system were laid out: a shift from an exploitative, capitalist, colonial worldview with its accompanying systems to an anti-extractive, anti-capitalist, anti-speciesist, and anti-racist society.

Within this reframing, many participants spoke directly to the need for a decolonized worldview that includes respect for and reciprocity with the land, with diverse groups of humans, and with other animals. An example of this positioning comes from our interview with Vandana Shiva, who told us: "So I think on everything, whether it is resilience or it is democracy or it is pluralism and possibilities ... diversity of age, diversity of species, diversity of values, diversity of intelligences, we've got to allow all of those diversities to flourish. And our work is to create the conditions of diversity" (Interview with Vandana Shiva 2023).

At the heart of any new paradigm should be transparency, one of the foundational pieces of building a truly democratic system.

Participants highlighted the need for transparency to facilitate change, but also pointed to its limitations: transparency alone might not be enough. Instead, some participants suggested combining transparency with broader processes of education to facilitate empathy and compassion for others. Some articulated the hope that as citizens gain a greater understanding of how others are affected by the unjust food system, and gain a better sense of the overall issues we face collectively in terms of climate change, they may get to a place where they can essentially put themselves in someone else's shoes, so to speak.

This, of course, requires gaining an appreciation of, and compassion for, many of the realities that participants described from their own work and research: the experiences of workers in this intensive food system, whether they be farmers struggling to make ends meet or slaughterhouse workers; other animals commodified and prohibited from living their best lives; diverse groups of humans forced off the land, or prohibited from growing their own food, or procuring food that is meaningful to their culture, and so on. These are just some of the examples of the kinds of stories we heard during our interviews.

The discussion of empathy and compassion was not only framed within the larger, more abstract sense of caring for populations of people and other animals, or ecosystems. Many participants emphasized the importance of valuing *individuals* within these diverse populations as well. When participants included other animals in their framing, and the overwhelming majority of participants did, they included notions of personhood, a recognition of other animals not only as part of species and ecosystems but also as individuals wanting to live a good life. It was broadly acknowledged that a fundamental aspect of shifting paradigms is recognizing that other animals and nature deserve consideration and compassion.

Community Building

"Community building" was a theme that emerged in nearly every interview. We have discussed the importance of strong and resilient communities, for both humans and nonhuman animals, not simply because this is a positive for communities in general as they improve their collective support system, but because it will become *essential* in these times of climate disruption. As we grapple with the enormity of the issues facing us through climate change, and the disruptions it is already creating, community remains an important way to insulate ourselves from some of the worst harms (Klinenberg 2018).

Several issues related to community building were discussed by participants and are worth noting here, as they point to potential solutions. In their descriptions of community, many participants reflected a belief in interconnectedness, coexistence, and oneness (the broadest sense of community). They highlighted the important connections between building community and ties to cultural traditions (e.g., culturally appropriate food as a way to counter colonial diets imposed on Indigenous communities). These discussions, it was noted, can facilitate a re-engagement with food, highlighting the powerful idea that cooking is political.

Some participants related strongly to the idea of "degrowth" and reflected on its implications both structurally (slowing down production/consumption so that it is within sustainable limits and also respects workers and other animals) and also individually because it can increase time for convivial relationship-building, time for other non-consumptive pursuits, and provide space for meaning and value outside of the consumer culture. Discussions about food can also provide agency to communities seeking to decide what food security means for them specifically. At the same time these discussions can address pressing structural issues related to the colonial/capitalist mindset and the economic relations that fracture us and create divisiveness.

In discussing activism and community building in the food movement, farmer Len Vassallo notes: "I think there's a hunger for connection ... There are all kinds of possibilities for working in a community setting. And I think that's one of the things we've lost in our society as it's gotten so individualistic that it's lost a lot of the sense of community. And [with community food initiatives] we gain our food, we gain our communities, we gain back our health, our physical health, our mental health" (Interview with Len Vassallo 2023).

Initiating Effective Change

The final theme that we want to bring in from our interviews relates to the question of how we initiate change. While the diverse participants used different analogies and perspectives to frame these ways forward, they had much in common: *mosaic, spider web, Two-Eyed Seeing, and multispecies ethnography* were all discussed.

When we interviewed Ashleigh Long, an Indigenous undergraduate student at Cape Breton University, and Research Assistant on this project, we were interested in her viewpoints for three main reasons: what was it like to be learning about all of this in an immersion type of environment (reviewing literature and checking transcriptions over several months)? How was she, as an undergraduate student (the main audience for this book), feeling about the material she was learning about? And, finally, as a young Indigenous woman: what was her personal connection and understanding to the main issues framed in the book?

In one question we asked her: The primary audience for the book will be undergraduate students, so how do we go about making it accessible and appealing to them? She began her answer by saying that she thought framing, and bringing together issues related to the land, humans, and other animals was essential. She went on to note:

> Demonstrating how all of these things are connected in one way or another is key. If you were to create a spider web, for example, and start in the middle with the topic of meat consumption, you can then begin to branch off from that in so many other directions. The web represents the underlying interconnections that this one area has to other topics, which confirms the connectedness between each of them and how one can ultimately affect the other.
>
> A big piece in discovering this perspective was my realization of how connected everything is. Every article I have read, I would think, "Oh my God!," and it would open up another world for me. Then of course you go down this rabbit hole and you know, on and on and on. But for me, I think that was the biggest influence, the connection piece of how all of these different topics are linked in complex ways to one another and can have a simultaneous impact on one another. Like, hey, this happens because of this, and this happens because this is occurring, and when you look at the bigger picture, you realize that it is one big web of connections.
>
> If you are able to discuss even one topic that one person is interested in and then connect and tie everything together, I think it'll help people in realizing that you cannot focus on one aspect of the web without the consideration and understanding of everything else it is connected to. (Interview with Ashleigh Long 2022)

On a similar note, Katharina Ameli, Coordinator of the Research Centre for Animal Welfare and a lecturer at the Institute for

Sociology at Justis-Liebig-University in Giessen, Germany, discussed her framing of the interconnected issues, and potential solutions, as a mosaic:

> In my opinion, it's always important to have an interdisciplinary view ... I'm working in an interdisciplinary way, so you can imagine it like a mosaic. You need different parts of the mosaic to get it together, and if I would have just a focus from a veterinary medicine perspective for example, [this is] just one mosaic stone. I could see maybe the animal, but from a sociological perspective, which would normally focus on humans ... I could have the mosaic stone for humans, but with the nutrition part, all this stuff I learned while I did the interdisciplinary incorporations that came into the mosaic. It is another mosaic stone, which has connections at the interface of humans and animals. And this is how I'm working. I'm working like a mosaic. (Interview with Katharina Ameli 2022)

Multispecies ethnography is another approach that opens doorways to both appreciation of interconnectedness with other species and to the wonders, intelligence, and complexity of the natural world. Katharina Ameli recently published the book *Multispecies Ethnography* (2022), which asks us to consider how we can integrate nature, humans, and animals, or what she calls "HumansAnimalsNaturesCultures," and form a more significant understanding of the major issues and inequalities we face. When we asked her about the connection between food justice and her work on multispecies ethnography she said:

> I think that this could be a good method to show all the parts about food. Or like I said at the beginning, if you eat something, if you would do a multispecies ethnography for example, with the plant or with an animal, you could try to analyze all the parts, the whole morphology of all [the] stops in between. Not just eating, it's more about growing, it's about how it came to you, [such as transportation]. If cooking, how do you cook it? Then eating, where's the plate [from]? So, there is a lot you could observe in this world with a multispecies ethnography, which is also complex, because if you see it from the complex side, you have to include so many different perspectives. (Interview with Katharina Ameli 2022)

Multispecies ethnography provides a tool to help us look beneath the surface to recognize the connections and complexities of all that is around us. It suggests the need for interdisciplinary and holistic approaches that recognize the agency of other animals and nature, while respecting, and weaving in Indigenous knowledge systems to help us grapple with the serious issues we face.

Indigenous knowledge systems in particular provide a foundational approach to exploring, critiquing and reimagining relationships with the land and other species. In trying to figure out what this reality means for each of us in our daily lives, it is important to acknowledge the places and spaces that we personally inhabit and work outwards from there. In this, we acknowledge our place here in "Nova Scotia" on unceded Mi'kmaw land and the perspectives of the Mi'kmaw People historically with regard to this land.

The Two-Eyed Seeing approach developed by participant and Mi'kmaw Elder Albert Marshall, the late Elder Murdena Marshall, and biologist Cheryl Bartlett, begins with the concept of "Netukulimk," which expresses the idea of living in harmony with the land but not in a subject/object way as the anthropocentric discourse of sustainability suggests (Bartlett et al. 2012). Rather this thinking is organically expressed in our ways of being on the land, with each other, and with other species, and is captured in the term "Msit No'kmaq," translated as "All My Relations." This phrase is meant to express the idea that all living beings are our relatives and thus deserve the ethical treatment we would extend to our family or kin.

Two-Eyed Seeing is often misconceived as simply based on the dialogue and co-learning between Western and Indigenous knowledge systems in a way that captures what is best in both traditions, allowing us to coexist healthfully together in the future. While this is a key part of the picture, Elder Albert Marshall clarifies that Two-Eyed Seeing is meant to incorporate multiple perspectives and acknowledges that other cultural perspectives from Asia, Africa, and the Middle East will also be key to determining emancipatory ways forward. Bartlett explains how the ideas developed: "Albert, a fluent speaker of Mi'kmaw, reflecting upon the wisdom within Mi'kmaw Knowledge, as held within the language, suggested the word 'Etuaptmumk' which translates as 'gift of multiple perspectives.' He pondered further and coined the English phrase 'Two-Eyed Seeing' as the 'traveling companion' (Bartlett's phrase) for Etuaptmumk" (Personal correspondence with Cheryl Bartlett 2023).

The concept then had the two components blended becoming "Etuaptmumk"/"Two-Eyed Seeing" (E/TES). She notes that recognizing "the gift of multiple perspectives" is likely common to many Indigenous Peoples. Bartlett further explains that there was an early emphasis on co-learning and co-respecting, applying both together. Marshall emphasizes the incredible importance of storytelling and sharing across cultures. The sharing of "knowledge stories," Barlett notes, is the way E/TES always begins positioning "Netukulimk" within Mi'kmaw Knowledge as the lead story from there. Western Science or Western perspectives may or may not contribute to the story, depending on the situation.

According to Marshall, it is in this process of relationship building through stories that we begin to understand where we came from, who we are and why we are here. For the Mi'kmaq, this process is grounded in the Mi'kmaw Seven Sacred Gifts of Life. These teachings are built around the prerequisites, values, and corresponding practices of Wisdom, Love, Truth, Bravery, Respect, Humility, and Honesty (LearnRidge 2023).

The idea of humility, translated from the Mi'kmaq word "Wanqwajite'teken," can apparently also mean compassion: "Humility is to know yourself as a sacred part of Creation" (LearnRidge 2023). As we have argued elsewhere, what some may see as a sacred connection between ourselves and all creation, a more secular person may view as the science of interdependence (Gibbs 2017). As Albert Einstein pointed out, nothing in the universe exists in isolation, as separate from other things; this is scientific fact. For the individual to view themselves as separate was to Einstein "a kind of optical delusion of his consciousness" (Einstein 2015). For us, the two perspectives complement one another and, in the spirit of Two-Eyed Seeing, effectively add up to the same thing. We believe that this orientation not only gives rise to an inherent compassion in us as humans but can inform the path to more just relations with all species and the land while complementing the Western idea of the precautionary principle which we will explore more fully below.

Most participants spent time exploring important ways to create spaces and places for respectful dialogue across social movements and diverse cultures. There was often a strong orientation of the importance of meeting people where they are, having democratic participation in decision-making at a local level, and, for those from the Global North, an insistence that any change start in correcting the oppressive practices and policies here at home that have implications on a global level. This was especially urgent for the industrial production of nonhuman animals as *food*, in terms of the implications for those nonhuman animals' lives, worker and citizen health and safety, and the immediate implications of the system for climate change.

Ideas about how these changes could be implemented varied widely – *reduce, refine,* or *replace* – and were discussed by some, whereas others urged more sweeping shifts such as plant-based diets needing to be more broadly instituted. Regardless, across the board, there was a strong agreement that the current industrialized food production system, and the ongoing utilization of billions of nonhuman animals each year, could not continue as is.

In thinking through the changes needed to this system and in consideration of the harms to nonhuman animals, some argued that incremental change, because it offers short-term easing of some of the "worst" harms, is sometimes necessary. No one argued that this was enough, and most argued that there is simply no *humane* way to *farm* nonhuman animals on the scale that we currently do. And some participants stated very clearly that "humane" is just the absence of cruelty; it does not provide for the necessities of a full, rich, or good life. Along a similar vein as thinking about incremental shifts, some participants indicated that in order for policy shifts to be effective, there must be adequate governmental oversight.

In thinking through some of the underlying solutions proposed and linking to both the need for a paradigm shift and in building community, many argued that we need localization while looking/thinking globally. For some, this related to overcoming supply-chain issues and some of the pandemic bottlenecks we discussed in chapter 3. For others it related to eliminating "plantation" agriculture so that communities could feed themselves or determine what they need to feed themselves. Still others discussed the need to reorganize the food system to deindustrialize agriculture, support small-scale farmers, and value farmers and food workers. This clearly links to the overall framing of the book, that food in a just world requires a new kind of democracy and social system.

Finally, if we want to continue to wake up every morning and feel able to work for change, we need to find healthful coping strategies. A wide range of emotional responses to the crises discussed in this book were identified in the interviews. Some discussed feelings of grief, trauma, and culture shock. But participants also identified positive feelings such as well-being and hope. Sometimes participants simply said that it was nice to talk about these issues, and the emotions they brought up, with folks who were also experiencing the same thing. There was a recognition that taking care of oneself, while simultaneously caring for others, was important to be able to continue to research, teach, learn, work, and advocate for change.

As we noted at the beginning of this book, all questions of food justice require that we first explore relationship to the land, both historically and in the present day. In previous chapters we have explored the ways in which these relationships have been systematically shaped and corrupted by colonialism and industrial capitalism and how the A-IC is just one manifestation of this broader system. From genocide and slavery to contemporary relationships of neocolonialism expressed in oppressive extractive activities and corporate domination of "democratic" systems globally, it becomes clear that without far-reaching changes, the human species and most other species face a bleak future.

With this possible future in mind – a future governed by unpredictable climate disruption, species decimation, and ecological breakdown – people around the world from community gardeners to land and water protectors are fighting for and showing the way towards alternatives. In these struggles, we see that many communities are achieving small and large victories and are paving the way for new ways of seeing and being.

Sorting Out the Politics: Short and Long Term

With Msit No'kmaq (All My Relations) and the precautionary principle in mind, we don't necessarily have to agree on how severe the climate crisis is, where it is headed, or on how to narrate the current situation to decide how to behave today. While it is important to raise arguments and to support these with facts and science, and to recognize the urgency of the crisis, it seems to us that too much time and intellectual and emotional energy is being used trying to convince others of "facts" and stories in the hope they will shift their thinking.

The kinds of values that we would like to see in the world – compassion, radical democracy, inclusivity, justice, and harmonious relations with the land and other sentient species – are, we believe, worth working for irrespective of all the debates around how we should best approach climate change. The good news is that we can incorporate these principles into our lives on a daily basis *right now*. At worst, these values will float unhinged in the incredible and beautiful experiments and ideas we see brewing around the world without managing to tip the scales of mass industry and commerce towards structural changes, and perhaps not working in time to avoid calamitous climate disruptions or a possible end of the human species in the future.

A middling perspective would argue that, at minimum, these ideas and practices will continue to spread like a good virus engaging more and more people while helping communities to build more resilience, to train in more democratic, horizontal processes, and to work with nature and other species in mind. From this angle it is well worth building our resilience and compassion as communities in order to be in a much stronger position to deal with the climate disruptions and concomitant social breakdowns that now appear inevitable (Jamail 2020; Bendell & Read 2021). At best these efforts may not only threaten the system, but over time they could tear down the institutions and processes that ensure the continuation of the experiences of structural violence against land, people, and other animals that we have identified throughout this book.

One of the challenges in promoting this broader agenda is that silo politics and sub-narratives have dominated the discourses of social justice and climate change. A lack of intersectionality in our approaches to social justice has meant that we have not properly understood the interconnections and interdependence between different social struggles and how these struggles relate collectively to the movements to protect nature and nonhuman animals (Crenshaw 1989; Gibbs 2017; LaBronx 2018). We have tried throughout this book to argue that intersectionality is a prerequisite to achieving social justice as a collective reality for humans and for our relationships with other species and the land.

In addition to intersectionality, structural analysis situating these various oppressions is also necessary. Aviva Chomsky explains this well in her analysis of the Green New Deal in the United States, which promotes the idea of green growth. She notes, "Green growth avoids confronting the resource use and pollution created by all forms of production, including that of so-called green or renewable energy. Even as green growth often advocates some forms of redistribution domestically, increasing corporate taxes and public services, for example, US versions tend to assume the country can continue its high levels of consumption and corporations their high levels of profit" (Chomsky 2022: 173). Chomsky does note that some supporters of the Green New Deal acknowledge that ongoing growth is actually impossible, and they advocate for slower development and better work–life balance as well as an openness to options beyond the market in our search for the better life.

Education professor Catherine O'Brien, pioneer of the concept of "sustainable happiness," argues that many of the practices that support us taking care of the planet can also actually make us happier overall as they reinforce family and community bonds, shared community practices, and more mindful localized consumption wherein we develop ongoing relationships with those who produce our food. This collective approach to happiness within a context of ecological sustainability has at its center the notion of "well-being for all" and underscores the fact that our happiness and well-being are interconnected with other people and the natural world (O'Brien 2016; Interview with O'Brien 2022).

In cooperation with schoolteachers, other educators, and graduate students, O'Brien has adapted her work for integration into the school curriculum, influencing classroom practices and increasing the emphasis on outdoor learning and engagement with nature. Plant-based meals reinforce the notion that one of the most powerful things we can do to protect the earth and our own bodies is to adopt a largely plant-based diet. As we noted earlier, these efforts are now complemented by the Canada Food Guide which in recent times emphasizes diets that are largely plant-based. This type of education is one antidote to what non-fiction writer and journalist Richard Louv calls "Nature Deficit Disorder." He explains:

> I coined the phrase to serve as a description of the human costs of alienation from nature and it is not meant to be a medical diagnosis (although perhaps it should be), but as a way to talk about an urgent problem that many of us knew was growing but had no language to describe it. The term caught on and is now a rallying cry for an international movement to connect children to the rest of nature. Since then, this New Nature Movement has broadened to include adults and whole communities. (Louv 2012)

The Art of Transparency

Transparency is also fundamental to radical democracy and is based upon a belief about honesty in communication. In other words, citizens have the right to know the truth about what is in their products and whether or not their products have been made in ways that compromise biodiversity, harm or threaten other species, or violate basic human rights or, in some cases, all of the above. As we have seen, the fact that in our current economy the onus is mostly on the consumer to figure this out helps conceal the structural violence inherent in the mass production and consumerism of the globalized economy.

The systemic – and sometimes directly physical – violence that is perpetuated through the various threads we have identified in this book as part of the "logic" of capitalism, reinforces many of the intersectional problems of racism, sexism, trans- and homophobia, and speciesism, allowing them to run rampant in the world. How did we arrive at a point in the so-called "progress" of our modern, supposedly democratic, societies that it isn't required, and is often actively avoided, to look at possible harm as a first principle in production?

If we translate this to other areas of life, we soon realize the absurdity. Most people would agree that it is wrong to cause violence to a child and would be offended if it was suggested that we should just be able to do this, and then after the fact people could then decide for themselves whether to engage with us or not, or choose to punish us or not, based on their individual choice, not based on some law or external force. And we would also think it absurd to rely on the perpetrator of the child violence to self-regulate their behavior and to come forth and admit their guilt and willingly change their ways.

Policymaking and regulation have the potential to make a difference if integrated with tangible governmental and judicial oversight. While debates still rage about their efficacy, designations by regulatory bodies and agencies based on ethical criteria have flourished in recent years, with labeling ranging from "organic" to "fair trade" to "ethically sourced."

Member of Parliament in Canada, Jaime Battiste, who serves as a member of the Liberal Party, made history by getting a sustainability labeling bill passed in the House of Commons. As he points out, "We've kind of realized that allowing businesses to self-regulate is not reliable. When people purchase something, they're buying into the practices of that company. Then I think that as long as there is no kind of label that's reliable, consumers are in the dark."

Battiste notes that most people are buying with their wallets in mind and not based on conscience, so a lack of transparency makes it more difficult to address the harms inherent in our current systems of production. Ultimately there is no immediate incentive to do the right thing. As he explains, "What my hope was, and my hope still is, when we looked at motion 35, is environmental labeling. We know that the industry is already kind of shifting to the plant-based meats and impossible burgers and sausages and ethical fair-trade coffee and things like that. We're starting to make that pivot as a society and trying to say that maybe we should be making better choices for our future and for the planet" (Interview with Jaime Battiste 2023).

While it is justifiable to be skeptical about labeling alone as a strategy to end structural violence, it could be seen as one piece of a much larger puzzle. Obviously, ideally, harm should be avoided and in the worst-case scenarios such as rights violations, it should be made *illegal* at the site of production. Despite this, we would argue that transparency in production and consumption, wherever it can be practiced, is still a good thing. And even to move towards suggesting that corporations should be transparent in their behavior, sadly, is a radical departure from business as usual.

Radical Democracy and the Precautionary Principle

To move beyond "tinkering with the system" we ultimately have to confront the backwards and inherently harmful approach to ensuring ethical standards practiced by so-called "civilized" modern societies. As Mahatma Gandhi said when asked what he thought about Western civilization, "I think it would be a good idea." Such a civilization requires radical democracy, which would be grounded in the principle that all activities, including economic activity, would be decided collectively prior to implementation.

Elder Albert Marshall has argued for an approach to all activities that we engage in as individuals and as societies that begins with the question: to what degree is what I am about to do in harmony with nature and other species? This idea, echoed in the Western concept of the precautionary principle, could frame all political and economic decisions made by communities and would be naturally included in what Kimmerer calls the "democracy of species" and what Vandana Shiva calls "Earth Democracy." Of course, it is inevitable that we will affect nature and other species in our activities as humans, but we can end the worst harms that are occurring and move closer to more harmonious relations with nature.

While we might be justified looking at the Scandinavian societies as the most advanced in the Global North in terms of social democracy (stellar social programs, housing, restorative justice, education, health care, and so on), Finnish scientists have noted that on the ecology front their score card is as low as other northern countries. At first this may seem baffling as these countries are known for their innovative investments and commitment to alternative technologies. But as we argued in chapter 2, a nation's ecological footprint is not simply based on what they do within their own borders, it is also about where and in what a country and its corporations invest abroad and what they import and consume. It is also important that these too are reflected in their ecological footprint.

As an independent Finnish multidisciplinary environment research unit BIOS notes in a recent report:

> The economy has been seen through a market economy lens, and the public sector has been viewed as dependent on the private sector. Therefore, economics have dictated political realities while facts about natural resources and ecological limits have had a minor role. However, this view is backwards. The market economy exists only as a part of the wider society and the wider society only as a part of the biosphere.

The report goes on to argue that technology alone (such as carbon capture and storage) cannot get us to the 1.5 degrees Celsius target, suggesting that "a thorough transformation of ways of life, economy and infrastructure is also necessary." Pointing towards an approach that considers this broader framework, they also note:

> The use of natural resources must be reduced to a level that is globally sustainable (roughly one third of the current average per capita consumption in Finland). More satisfactory and meaningful lives are possible even with lower material consumption. It is possible to organise the reduced use of natural resources in ways that increase equality and democracy. All Finns must be guaranteed a sufficient level of material well-being. (BIOS 2019)

As arguments about ecology-versus-economy continue to dominate the policy discourse around climate change, leading to an inability of countries to seriously confront the crisis, it becomes clear that such a precautionary principle may be better placed in a constitutional document (as countries such as Ecuador and Bolivia have done).

Clearly any vision of a just and compassionate food future depends upon our ability to work with and adapt to climate disruption; there is no way around this reality. This is the chaotic framework we are working within, but as the report above acknowledges, it will require creating the necessary social infrastructure, not only because it is ethically a good idea, but also because it will allow us to be more resilient and compassionate in the ways we deal with the climate crisis.

We began this book narrating our region's response to Hurricane Fiona, and, as we worked on this final chapter, an "historic" cold spell with severe winds hit our region, freezing water pipes and shutting power down. As we have made clear, how the effects of climate change and the disruptions it causes play out varies across communities, regions, and countries, and also unequally across race, class, gender, and species lines. With this reality in mind, the modest agenda of "green economics" and the narrative of sustainability, which have clumsily coexisted with the ongoing logic of capital, are clearly not enough.

Resilience, compassion, ecological harmony, and justice will require an intersectional approach that is rooted in grassroots communities with meaningful control over their destinies. Elder Marshall notes that self-sufficiency may not always be possible (nor necessarily desirable); we will have to trade and engage beyond our communities, but self-reliance may be possible where a genuine and radical democracy is nurtured.

The Importance of Joy and Gratitude

While it might sound trite to include joy and gratitude in a very practical political book on food justice in a time of climate change, we would argue that celebrating the beauty of nature and feeling a sense of gratitude for the other species we share this planet with is a fundamental part of storytelling and community building. Both not only enrich our lives and contribute to resilience, but they give very real meaning to the struggles for social justice and ecological harmony. We can and should avoid harm wherever possible and to do so requires a sense of appreciation for all of nature and other living beings.

Like many who work closely with other animals, president and co-founder of Farm Sanctuary Gene Baur speaks of "mutuality," suggesting that when we harm or cause violence towards other beings, we break the bonds of mutuality. As he explains:

> I've really been thinking a lot in terms of creating systems of mutuality as opposed to those of extraction. In the dairy industry, for example, you're extracting milk from cows, and taking their babies away, and then when they are no longer profitable on the dairy, they're killed for meat, and ultimately, their lives are taken away from them. To raise animals for food, you destroy ecosystems and extract nutrients from the soil, and factory farms also extract taxpayer dollars from the government. So this is a highly extractive system that is the opposite of mutual. (Interview with Gene Baur 2022)

Baur goes on to acknowledge that many see the products of non-mutuality, such as a *hamburger*, as some kind of right and this sense of entitlement is reinforced by the industry. And, as he notes, beyond other animals, this sense of entitlement permeates many of our food choices in the Global North, from feeling like we should have access to fruits and vegetables of all sorts all year round irrespective of the possible costs to nature and/or the disempowerment of communities growing the food.

According to Baur, the way forward is definitely not to be "the political correctness police," who are never welcome and rarely achieve their goals in social situations. Instead, we need to find places and spaces where conversations can take place without judgment, and this is a critical piece of building a shared culture around mutuality. Nobody wants to hear that they are doing something bad or cruel and, if they feel that judgment, they will most likely respond defensively, reiterating the accepted narrative that this is what humans have always done, it is part of our nature, and that *meat* eating is what "normal" people do.

What we have argued throughout this book is that the ways in which we produce most of the *meat* and *dairy* and *eggs* and *fishes* for *food* is quite new historically speaking, and the cruelty involved is not "normal" or healthy, either for the nonhuman animals themselves or the humans killing and/or eating them. We have seen the mental health effects of slaughtering and *meatpacking* on workers as just one example. And, we have already noted that those curious about *humane* farming and plant-based eating as well as those reducing their *meat* and *dairy* intake are important allies in the drive towards more compassionate food systems. Also, numerically speaking, they form a significantly larger portion of the global population than vegetarians or vegans.

We can return to the themes of joy, appreciation, and acknowledgment of the emotional lives of other animals to see how a relationship of mutuality with nature and other species can be nurtured. This is a theme that has flowed throughout our research and in the writings and reflections of the authors and interview participants with whom we have engaged. When speaking of his own grief about what is happening to our planet, journalist and author of *The End of Ice: Bearing witness and finding meaning in the path of climate disruption* Dahr Jamail returns to his personal sense of awe:

> I find my deepest conviction and connection to the earth by communing with the mountains. I moved to Colorado and lived among them when I was in my early twenties, and it was there I began to deepen my relationship with them and began to really listen to them. I would hike out and just sit among the peaks, watching them for hours, and write about them in my journal. Today, I know in my bones, my job is to learn to listen to them ever more deeply, and share what they are telling us with those who are also listening. (Jamail 2020: 224)

Similarly, Robin Wall Kimmerer's *Democracy of Species* begins with a deep sense of awe towards nature and, in the vein of Two-Eyed Seeing, an appreciation for the Western science that opened the door for her to fully explore that world from another angle.

For those working directly with other animals, stories of animal beauty and expressions of joy at being alive abound. This is evident in any David Attenborough documentary, but it is also brought poignantly to life by individuals such as Jane Goodall, Marc Bekoff, Gene Baur, and Brandon Keim, who speak of their witnessing and engagement in the lives of individual animals. It is not possible to truly convey the power of their individual stories without engaging and witnessing the beauty for ourselves, but their stories reinforce what that engagement can do to transform our value systems and help us to develop more compassionate ways of being in the world.

In one story, journalist Brandon Keim told us about a community of people who have coalesced around the coyotes of San Francisco. He points out that these people:

> are really helping to create a culture of coexistence and really doing the work that is necessary to do that, which in a lot of cases means just talking to people who they meet while they're out. You know, so they're

going to the same park every day, they know the people they're talking to, they're not just like walking up to them delivering coyote messages. These are their friends and neighbors and acquaintances who they can talk to about coyotes and, through a million little conversations like that, they can kind of shift the way that people see these animals. (Interview with Brandon Keim 2022)

Similarly, Marc Bekoff's work on 'rewilding our hearts' is all about engaging with the world in a way that allows us to reconnect with our sense of wonder and interdependence with all species and with the earth itself (Bekoff 2014).

When Farm Sanctuary founder Gene Baur was asked about joy and what it may mean for other species, he responded by telling the story of Opie, the calf he rescued from almost certain death. We shared this story in chapter 3 but for our purposes here it is worth recalling a larger issue raised in the story about how animals such as cows thrive in relationships with their own species. At first, after the rescue, Opie was doing all right physically, but wasn't thriving until Baur realized that what he needed was to be "with his people." He notes, "so I brought him out to the cow barn and the cows joined around him and he just perked up and that was what he needed. So, Community is important too" (Interview with Gene Baur 2022).

We have argued in this book that recognizing that other species feel pain is important, to acknowledge suffering of any living being is a starting place for changing our behavior towards them. We accept this is true of our companion animals but what we have found in our conversations with many people about nonhuman animals considered to be *food* is that once they realize the conversation is not going in a judgmental direction, they will open up more and express that they often engage in "blocking out" what they know to be true.

Similar to Gillespie's use of "doublethink," or Arluke and Sander's use of "ambivalence," "blocking out" is reinforced by the food industry through language and advertising, so that we do not have to deal with those lingering uncomfortable feelings. But knowing that other animals like to play, can feel joy, and can also feel a sense of community and, some would argue, even a sense of culture, can and does shift perspectives.

But there are few places in our society to connect with these stories; one has to seek them out and those who do are generally part of the already "converted." Some believe that our best bet in raising awareness around harm to other animals in a consumer society is through policymaking, regulation, and labeling. But Sophie Riley notes that anti-cruelty regulation and animal *welfare* can only go so far as there

> has been an inequitable bias in favour of human interests, where a pragmatic approach has led to farm animals being treated as commodities. This has led to the entrenchment of commercial biases because society has long accepted the fact that animals and their products are goods to be traded. Moreover, these assumptions accept the validity of animals' classification as property and the legitimacy of using animals in an increasingly intensified way. This situation is unlikely to change without government exercising its political will and power to correct imbalances in the sector. Only then will society evince a view of farm animals as living beings worthy of protection, rather than seeing them as commodities to be exploited. (Riley 2022: 215)

We have spoken in this book about denial being the first stage of grief. Blocking out what is most oppressive about our systems of food production and consumption, indeed of our consumptive lives in general, is now a practice so deeply entrenched that it can be hard to see our way out. This cognitive dissonance is what allows us to continue supporting structural violence even when at some fundamental level we may recognize it as a problem.

Many of the practices mentioned in this chapter that involve engagement with nature and other animals, practices of gratitude and appreciation, can serve as powerful antidotes to our collective denial, allowing us to connect in an immediate way with our sense of kinship with all other animals and living things. This happens much in the same way that we connect with and can appreciate other cultures by actually sharing stories with them and learning from them. This is something that many of our participants such as Elder Marshall emphasized as fundamental to how we can transform our societies away from many of the "-isms" that separate us. In addition to this, we have argued in this book that demanding transparency and building it into everything from our daily practices, institutional structures, and policymaking is a big part of the solution.

If cruelty and ecological destruction or rights violations are not directly visible to us, they are easy to ignore. In the language of psychology, blocking out is a human instinct, one that protects us from harm or insulates us from the effects of trauma; it is how we have taken care of ourselves in the face of danger since our early existence on this planet. But psychology also notes that while this blocking out makes sense in a situation where our survival depends on it, it can lead to dysfunctional behavior and even mental illness when the danger is no longer present or is not in fact real.

The adage that knowledge is power may seem trite, but our capacity to live life fully and to ensure that we and other species can live lives of joy in a diverse natural environment requires that we face the truth with humility and compassion.

Conclusion: Toward the Compassionate World

In writing this book we have acknowledged that ours is just one lens on the multitude of issues raised as we aspire to face with compassion the urgency of our current global crisis of ecology and justice and how this affects our food systems. For this reason, it is with humility that we offer a "conclusion."

We have tried to strike a balance between making clear that much, if not most, of the world lives in really difficult economic circumstances and, as Oxfam and other institutions make clear in their research, many live in "obscene" poverty and hardship. And the suffering of other species, and that of so-called *food* animals, is inextricably linked to the systems that maintain this poverty and inequality for humans. In terms of humanity, a majority of the millions facing malnutrition and famine today are disproportionately racialized people in the Global South.

Actions carried out by Global North countries tend to be reflected in contributions to aid programs through the United Nations rather than confronting the unsustainable and unjust economic model that is killing people daily, again, mostly in the Global South. Food insecurity in many countries is already at a crisis level and, if the predictions of climate scientists are accurate, this crisis could hit the Global North in the not-too-distant future (Jamail 2020).

These issues, in our view, are directly connected to colonialism and to the broader history and contemporary manifestations of the capitalist system which is governed by its own logic. At the same time, we have tried to highlight the fact that many people and communities from the Global South are leading the way to a better future through their activism and through their living practices of alternatives. We have argued these examples also contribute to the possibility of better lives for our nonhuman relatives and for the rest of nature. In terms of trying to save the Amazon alone, often called the "lungs of the earth," the activism of Indigenous land and water defenders, on behalf of the rest of us, is worthy of nothing less than awe. In Brazil, activists are being assassinated at an alarming rate every week.

We began this book looking at the significance of our relationship to the land and our relationships on the land, tragically demonstrated under our current economic system in the countless deaths of humans and other animals. Contemplating land as public domain, as a "commons," to be used for the good of all is an ancient idea and one that many theorists, activists, and scientists are trying to resuscitate. "Protected" areas and "endangered species" policies are ways under capitalism that we have tried to protect heritage and biodiversity, but the climate crisis is making it clear that this is not enough. What if all land were to be a "commons" with people managing their individual dwellings and land with a requirement that nonhuman animals and biodiversity are protected as best as possible?

We have tried throughout this book, and through the voices of our participants, to highlight inspiring examples of this kind of thinking from around the world. Navdanya in India is one example where an impressive network has been developed to empower small farmers and producers to reclaim the land, for the purposes of moving towards what Vandana Shiva calls "food sovereignty." As Shiva told us, Navdanya is primarily a "seed saving, seed freedom" movement that facilitated the creation of 150 community seed banks. They have trained more than two million farmers in seed sovereignty, food sovereignty, and regenerative organic agriculture. Their work has shown that both health per acre and wealth per acre increase when farmers intensify biodiversity instead of chemicals (Interview with Vandana Shiva 2023).

We have applied this framing of food production to the context of radically democratic systems that are by nature and structure more compassionate. Many specific countries such as Cuba, Sri Lanka, and Costa Rica, to name just a few, have made impressive efforts to build more genuinely "sustainable" societies along many of the lines identified in this book. In practically every country in the Global South where the authors have either worked, visited, or conducted research, there are incredible examples of resistance and alternatives, and have been for years.

For this reason, our teaching, research, and activism over the past 30 years have led to increasing humility. It has also been made clear that many experiments or alternatives that could move our world to more just systems (in which food justice would be taken for granted), sadly, have been crushed in global power struggles and neocolonial interventions.

The "methods" of intervention are varied and include the economic measures of the international financial institutions such as the World Trade Organization, the International Monetary Fund, and the World Bank; the policies of international trade; and the practices of global corporations. Additionally, many analysts and a couple of our participants highlighted other forms of intervention by Global North nations, particularly the United States. Examples include the blacklisting and persecution of activists and community members; physical violence against those opposing extractivist, speciesist, and ecologically destructive projects; interfering in elections; and the financing and arming of "freedom fighters," militaries, and paramilitaries (Jamail 2020; Interview with Ceallaigh S. MacCath-Moran 2022; Interview with Aviva Chomsky 2022).

We have argued that building a just food system requires just and radically democratic societies that are transparent in their relationships with other communities and nations. Consequently, it is crucial that we expose the coercive and often covert interventions mentioned above to highlight how democracy is undermined in Global South nations. Making these types of interventions visible is essential for achieving decolonization, justice, and radical democracy in countries of the Global South.

We also conclude by acknowledging a bias in our lens as we focus on messages about decolonization, intersectionality, and social transformation. While we hope that our messages are supportive and entirely collaborative towards those around the world building alternatives, our focus for change is the people and institutions with privilege and power who support the systems of structural violence we have identified in this book. But it is not about identifying bad and good people. As we have acknowledged, many average citizens in the Global North are simply unaware of the systems perpetuated by passive acquiescence to business as usual. But our silence, whatever the reason for it – choice or ignorance – as Elder Marshall warns, allows the status quo to continue.

Both authors, in their research, teaching, and work with communities and organizations in the Global South, or with marginalized sectors in both the North and South, have focused on what those of us in the West/North/privileged places (and with privileged identities) could or should be doing to get our own act together. Let the experiments and alternatives moving the world towards a more compassionate and just order arise wherever and whenever they emerge, and learn from and support them in ways that they have identified as important to them.

Structural violence is a consequence of a capitalist system that has also caused the climate crisis we find ourselves in and in which industrial food production is just one thread. But, as we have argued, it is an absolutely vital thread. Literally, it is a threat to the water we drink and the food that we eat in order to survive. With this in mind, we aspire to work with our human and nonhuman relatives all over the world who are living on the land sustainably and who are working towards creating a peaceful, radically democratic, and compassionate world. This approach could guide us to more liberatory ways of doing "development" and of creating "prosperity" in the spirit of co-learning as identified by Elder Marshall.

We began this book with a quote by Vandana Shiva, and it is here that we would like to return: "the violation of ecological limits is the beginning of ecological injustice" (Interview with Vandana Shiva 2023). So many of the crises examined throughout this book have focused on the pushing of limits – ecological, human, and other animals' rights – and many of those in control of, and benefiting from, the structures and processes that support the status quo claim that change will negatively affect our quality of life. But what if we reframe this so that, rather than focusing on what might be lost, we consider what we all stand to gain: ecological balance, rich community, interdependence, true democracy, and compassion for all.

Methods

In this book, we were interested in learning how people define food justice and resilience and how they are building resilience into their research, work, production, and/or advocacy. An assumption driving the qualitative research aspect of this project was that food, the sustenance of life, cannot be governed by the "logic of capital" if we hope to solve the related crises of systemic injustice and environmental collapse.

We knew from the beginning that we would explore five major topic areas related to industrial production of nonhuman animals as *food*: nonhuman animals, consumer-citizens, workers, the environment, and potential solutions. These broad themes can also be seen later on in the discussion of the coding process as they acted as *a priori* codes, to help us link to the emergent codes that arose from a closer examination of the transcriptions and that eventually help build a cohesive understanding of the interview data (Maxwell 2005: 96–8; Tilley 2016: 152–6).

This research was approved by two ethics granting bodies: the Mi'kmaw Ethics Watch (MEW) and the Cape Breton University (CBU) Research Ethics Board. We received funding to hire student research assistants through the CBU School of Arts and Social Sciences, the Office of Research and Graduate Studies, and the Cape Breton University Work Study Program. Research Assistants Ashleigh Long, Jenna MacNeil, and Rochelle Roach all worked on the collection of relevant literature. Ashleigh Long worked on the book project from beginning to end and was responsible for correcting transcriptions and citation formatting.

In addition to the use of autoethnographic methods, the study relies on in-depth, semi-structured interviews with key experts on building compassionate food systems for people, other animals, and the environment. The participants were carefully chosen for inclusion in this project. Key participants were recruited based on personal contacts we have or knew of because of our research, international solidarity work, and/or community service activities. Some participants mentioned someone else in their interview, or recommended someone else they thought we should speak to because of relevant experience and/or knowledge on a theme that arose in their interview. The recruitment strategy was a combined purposeful selection (Maxwell 2005: 89–90), with a modified snowball sampling.

First, we made a list of areas we needed to know more about and/or people we knew or had read about doing work in the areas being addressed in this project. We contacted those people to see if they would be interested in participating. During these initial contacts, we emailed potential participants a brief overview of the project and a list of general areas to be covered in the interview. Once we had arranged a time and date for the interview, we forwarded the verbal consent script so that participants could read through and ask questions prior to the interview. We also often forwarded the broad interview questions so that participants had the opportunity to think about the areas being covered ahead of time.

Prior to the interview, we secured verbal consent, asking if participants were willing to participate, if they agreed to audio recording, and if they wished to use their name or maintain confidentiality. All participants in the study wished to waive confidentiality. This relates to the fact that participants were being interviewed about their relevant life histories, work, research specialization, and/or advocacy work related to building a just food system.

In addition, because we wanted to ensure that participants had a say in how their words were used and how we introduced them in the book, we asked that participants read through their in-text quotes prior to publication. While this added another layer of complexity (and time) to the research process, we believe that it is also more respectful (Kovach 2009) and gives important validity and credibility to the research by ensuring that the transcriptions were done correctly and that the participants had a final say over how their words were used (Maxwell 2005: 111; Tilley 2016: 145–8).

The interview questions were purposely kept semi-structured and open-ended. We worked to facilitate a conversational feel to the interviews. As part of the purposeful selection of interview participants, and because of the emancipatory goals of this research, we also recognized early on that while we wanted to make room for a wide variety of viewpoints related to the broad area of creating a just food system, we did not intend to give room to viewpoints that simply reinforced the mainstream dominant paradigm of exploitation of the land, humans, and other animals. This is important because "social justice inquiry also includes taking a critical stance toward social structure and processes that shape individual and collective life" (Chamaz, as found in Tilley 2016: 38).

The interviews took place over a ten-month period beginning in the spring of 2022. Due to geographical constraints, the majority of the interviews were conducted via video on Microsoft Teams or Zoom. Three participants were interviewed in-person and one participant requested to be able to answer the interview questions in written form and they then emailed the completed transcript. In total, 28 participants were interviewed in this project, and Mi'kmaw Elder Albert Marshall completed a series of life history narratives that informed many aspects of this work, and to whom credit is assigned throughout the book.

Participants came from diverse geographical locations: Australia, Canada, Germany, India, Japan, and the United States. They also came from diverse cultural, ethnic, and racial backgrounds. A wide range of ages is also represented. Most interviews lasted approximately one hour but several lasted significantly longer. While the questions were tailored to our participants' experiences and expertise, there were several questions in common for all participants and these allowed the coding on those questions to be compared and contrasted for all participants.

While we asked that participants review their in-text transcriptions prior to publication, for Mi'kmaw participants we reiterated prior to the interview in the consent script, and again after, that they would have the opportunity to see which excerpts we intended to use, and the associated context and analysis. The Mi'kmaw Ethics Watch application process asks researchers to clearly consider the risks associated with the research project, not only individual risks but also risks to the collective. For instance, Question 4.1 of the application states: "Indicate if any aspects of the study involve risk to the participants or to the Mi'kmaw People collectively. Describe any risk to the person/persons as a result of the findings being reported or published or risk to the Mi'kmaq, such as to their treaty or Aboriginal rights." While the research questions did not pertain to Aboriginal rights specifically, there was the potential that a participant might tell us something that pertains to collective rights.

As all Indigenous participants had the opportunity to review their contribution prior to publication, they then had the opportunity to make changes and corrections prior to the final manuscript being sent to the publisher. Participants reviewing their quotes within the context that they appear in the manuscript, and being able to add clarification or further context, adds an important degree of validity to the analysis, as we had verification from interviewees that how we framed the analysis fit with what they had told us in their interview. It is also, importantly, more respectful and reciprocal (Kovach 2009).

After the interviews were completed, they were transcribed using the transcription software in Microsoft Word. Interviews were double-checked by research assistant Ashleigh Long, by listening to the audio file and comparing it to the dictation. Transcripts of the interviews with Mi'kmaw participants were reviewed by the authors as outlined in our Mi'kmaw Ethics Watch application.

We did not correct grammar, pauses, word stumbles, etc., during the transcription. Prior to using any quotes within the book, we used the *Transcribing, Editing, and Processing Guidelines* from the Minnesota Historical Society Oral History Office to help guide the final editing process. We were careful to only lightly edit as we wanted the "participants' 'voices' to be evident in their quotes" but we "also recognize that participants have a desire and deserve to have their words read clearly and intelligently" (Harris 2018: 102–3). Therefore, we also edited for "false starts," "stumbles," "extraneous remarks," and "slurred words" (such as "gonna" becomes "going to") (Minnesota Historical Society).

Throughout the interview process we worked with the interview data. Initially, and prior to the completion of all interviews, this involved engaging with the transcriptions and highlighting sections and/or making note of places in the transcriptions where the participant discussed something that seemed particularly relevant to the broad research areas and/or the major questions. Once the transcriptions were completed, we analyzed the interview data more holistically. We did this by taking time to further familiarize ourselves with the interviews, looking for common key words or themes that emerged.

After the initial pass through using open coding, we examined the transcripts again, using a more focused approach where we began to collapse some of the open codes, and focused on thematic connections between some of the early codes (van den Hoonaard & van den Scott 2022: 176–80). Because of the broad experiences and expertise of participants, coding was not a straightforward comparison between participants. We did think through the ways in which participants were addressing one or more of the major areas of the book, and this did allow us to find commonalities and differences across a diverse grouping of participants.

All 28 participants are quoted within the book, although not all participants are quoted as frequently. Each interview helped to shape the book and we are grateful for the significant contribution that each participant made. As we read through the interview transcripts, and completed book, we are reminded of the richness of in-depth interviews, and the stories told, and what a significant form of sharing they create.

Appendix

Research Participant Biographies

Katharina Ameli
Dr Katharina Ameli is an interdisciplinary scholar, holding a BSc degree and an MSc in Ecotrophology (nutrition, household management, and economics), and an education degree for German vocational teachers in subjects including nutrition, domestic sciences, and biology. She holds a PhD in sociology from Justus Liebig University (JLU) in Giessen, Germany.

Katharina is currently working as a coordinator at the Interdisciplinary Centre for Animal Welfare Research and 3R at JLU. Her teaching and research focus on the analysis of animal welfare indicators and professionalization in animal-assisted services within a broader analysis of a Culture of Care in animal sciences. She is the author of *Multispecies Ethnography* (Lexington Books, 2022), which proposes a holistic analysis of the human and more-than-human world.

Jaime Battiste
Jaime Battiste currently serves as Parliamentary Secretary to the Minister of Crown-Indigenous Relations and was first elected as the Member of Parliament for Sydney–Victoria, Cape Breton, in 2019. He is the first Mi'kmaq to be elected to Parliament. Mr Battiste has served on various parliamentary committees, including the House of Commons Standing Committee on Indigenous and Northern Affairs, where he helped review, examine, and report on the issues affecting First Nations, Inuit and Métis Peoples, and Northerners.

Mr Battiste previously held positions as a university professor, Treaty Education Lead, and Assembly of First Nations (AFN) Regional

Chief. In 2005, the National Aboriginal Health Organization named him as one of the "National Aboriginal Role Models in Canada." In 2006, as the Chair of the AFN's Youth Council, he was one of the founding members of the Mi'kmaq Maliseet Atlantic Youth Council (MMAYC), an organization that represents and advocates for Mi'kmaq and Maliseet youth within Atlantic Canada. He holds a Juris Doctorate from the Schulich School of Law at Dalhousie University and an undergraduate degree in Mi'kmaq Studies from Cape Breton University.

Gene Baur
A pioneer in the field of undercover investigations and farm animal rescue, Gene has visited hundreds of farms, stockyards, and slaughterhouses, documenting the deplorable conditions. His pictures and videos exposing factory-farming cruelties have aired nationally and internationally, educating millions about the plight of modern farm animals, and his rescue work inspired an international farm sanctuary movement.

Gene was instrumental in passing the first US laws prohibiting inhumane animal confinement and continues working on systemic food industry reforms. His work has been covered by major media outlets including ABC, NBC, CBS, Fox, *The New York Times*, *LA Times*, and *Wall Street Journal*, among others.

Gene has been hailed as "the conscience of the food movement" by *Time* magazine and named one of Oprah Winfrey's SuperSoul 100 Givers.

Marc Bekoff
Marc is Professor Emeritus of Ecology and Evolutionary Biology at the University of Colorado, Boulder, a Fellow of the Animal Behavior Society, and a past Guggenheim Fellow. In 2000 he was awarded the Exemplar Award from the Animal Behavior Society for major long-term contributions to the field of animal behavior. Marc is co-chair of the Ethics Committee of the Jane Goodall Institute and in 2009 he was presented with the St Francis of Assisi Award by the Auckland (New Zealand) SPCA. His latest books are *Canine Confidential: Why Dogs Do What They Do*, *Unleashing Your Dog: A Field Guide to Giving Your Canine Companion the Best Life Possible*, *A Dog's World: Imagining the Lives of Dogs in a World Without Humans*, and *Dogs Demystified: An A-to-Z Guide to All Things Canine*. He also

publishes regularly for *Psychology Today*. In 1986 Marc won the Master's age-graded Tour de France.

Aviva Chomsky

Aviva Chomsky is Professor of History and coordinator of Latin American Studies at Salem State University in Massachusetts. She has published widely on labor history, immigration and 'undocumentedness', Central America, Cuba, and Colombia. Her most recent books include *Is Science Enough? Forty Critical Questions about Climate Justice*, *Central America's Forgotten History: Revolution, Violence, and the Roots of Migration*, and *Organizing for Power: Building a Twenty-First-Century Labor Movement in Boston*, the latter co-edited with Steve Striffler. She has been active in Latin America solidarity and immigrants' rights movements for several decades.

Mark DeVries

Mark produced and directed the award-winning documentary *Speciesism: The Movie*, which screened at theaters worldwide and has been featured in Scientific American ("brilliant and compelling"), *The Huffington Post* ("tremendously entertaining"), CNN Headline News, *Psychology Today*, and *The Sydney Morning Herald*, among many others. Mark also filmed the world's first aerial drone footage of factory farms, released as part of his *Factory Farm Drone Project*, which has been viewed by tens of millions worldwide and received global press coverage. Mark has given talks at Harvard Law School, Stanford Law School, Dartmouth College, Brown University, and many other venues worldwide. Mark is also an attorney, licensed to practice law in Washington, DC.

Amy Fitzgerald

Dr Amy Fitzgerald is Full Professor of Criminology in the Department of Sociology and Criminology, and holds a hybrid appointment with the Great Lakes Institute for Environmental Research, at the University of Windsor. She is the 2023–2025 University of Windsor Vice-President, Research and Innovation Research Chair. Her research focuses on the intersection of harms (criminal and otherwise) perpetrated against people, nonhuman animals, and the environment. She is currently working on three grant-funded projects, and has published many peer-reviewed articles, chapters, and books. Recent book publications include *Animal Advocacy and Environmentalism:*

Understanding and Bridging the Divide (Polity Press, 2019) and *The Animals Reader: The Essential Classic and Contemporary Writings* (2nd edn, Routledge, 2021; co-edited with Linda Kalof). Amy is a founding member of the Animal and Interpersonal Abuse Research Group, the recipient of a Distinguished Scholarship Award from the Animals and Society section of the American Sociological Association, the Mid-Career Outstanding Faculty Research Award from the University of Windsor, and was a visiting research fellow in the Animal Law and Policy Program at Harvard University in 2020.

Owen Gibbs Leech
Owen is a student in the Bachelor of Science program at Cape Breton University in Unama'ki/Cape Breton majoring in Biology and minoring in History. Since childhood, Owen has had a fascination with the relationships and connections between humans and other species and the connection of both to all living things. A self-taught Citizen Scientist, Owen's years of homeschooling have given him the opportunity to read and research about all things living. Owen was interviewed for the publication *Living Schools: Transforming Education* (Catherine O'Brien & Patrick Howards eds., 2020). He is a member of the Climate Change Task Force Unama'ki/Cape Breton.

Jessica Greenebaum
Jessica is Professor of Sociology at Central Connecticut State University. Her research and teaching areas focus on the intersections of gender, feminism, animals and society, and vegan sociology. She has published over a dozen peer-reviewed journal articles and is actively involved in diversity, equity, and inclusion efforts on campus.

Michael Haedicke
Michael Haedicke is a public sociologist and Associate Professor of Sociology at the University of Maine. His research develops a critical perspective on food systems and environmental management practices. He is the author of *Organizing Organic: Conflict and Compromise in an Emerging Market* (Stanford University Press, 2016) and numerous articles and book chapters focusing on organic foods and agriculture and on the politics of coastal climate adaptation. Michael is also a member of the Maine chapter of the Scholars Strategy Network and the Senator George J. Mitchell Center for Sustainability Solutions at the University of Maine. His writings

about food and environmental issues have appeared in a variety of national and regional news sources.

Breeze Harper
Dr Harper has a PhD in the social sciences, with emphasis on intersectionality, anti-racism, and racial-gender inclusion/equity. She holds an MA in Educational Technologies (emphasis in racial-gender inclusion/equity in technology) from Harvard University, where she received the Dean's Award for her Master's thesis work on how racial-gender privilege operates in cyberspace forums. She earned her BA in feminist geography from Dartmouth College and received the Innovative Thesis award for her work on heterosexism in rural geographies. She has more than 15 years' career experience as a diversity, equity, and inclusion expert, ranging from curriculum development, to conference planning, to research and reporting, to publishing books and articles, to workshop design and facilitation, to recruitment and retainment, to critical content editing, and to strategic consulting.

Taichi Inoue
Taichi is an independent scholar-activist in Japan. Since 2015, he has translated works of Critical Animal Studies, which include *Animal Oppression and Human Violence: Domesecration, Capitalism, and Global Conflict* by David A. Nibert (2013), *Constructing Ecoterrorism: Capitalism, Speciesism and Animal Rights* by John Sorenson (2016), *The War against Animals* by Dinesh Joseph Wadiwel (2015), and *Animal Resistance in the Global Capitalist Era* by Sarat Colling (2020). In 2022, he published *The Forefront of Animal Ethics: What Is Critical Animal Studies?* – a comprehensive introduction to the field. His current research focus is on the relationship and solidarity between animal liberation and feminism.

Lynn Kavanagh
Lynn Kavanagh has an MSc in Animal Behaviour and Welfare from the University of Guelph and currently works as the Farming Campaign Manager at World Animal Protection, an international animal welfare charity with offices in 14 countries.

She oversees World Animal Protection's farming campaign work in Canada to change government legislation, corporate policies, and people's behavior to improve animal treatment and create stronger protections for animals. Lynn has been working in animal advocacy for over 20 years, beginning with a small grassroots organization which, at the time, was one of the few voices for farm animals in Canada.

Brandon Keim

Brandon is an independent journalist specializing in animals, nature, and science. His first book, *The Eye of the Sandpiper*, was published in 2017 by Cornell University Press. He has written three issue-length treatments for *National Geographic*: "Inside Animal Minds," "Secrets of Animal Communication," and "The Genius of Dogs." Brandon's work has also appeared in publications including *The New York Times*, *The Atlantic*, *WIRED*, *Nautilus*, and *The Guardian*. He has made broadcast appearances on NPR's *Science Friday* and *Here & Now*, PRI's *The World* and CBC's *As It Happens*. His book *Meet the Neighbors*, about what animal personhood – knowing them as thinking, feeling beings – means for our relationships to wild animals and to nature, is forthcoming by W.W. Norton & Company in 2024.

Richard Keshen

Richard spent his career teaching Philosophy at Cape Breton University in Nova Scotia. He is the author of *Reasonable Self-Esteem* (2nd edn, McGill-Queens, 2017). He is currently working on a book on Canadian political philosophy. He was a graduate student in the early 1970s at Oxford University where he and other students such as Stan Godlovitch, Ros Godlovitch, and Peter and Renata Singer helped to lay the philosophical foundations for the animal rights movement. He is a founding member of the Animal Ethics Project at Cape Breton University.

Ashleigh Long

Ashleigh is a recent alumnus of Cape Breton University, graduating with a Bachelor of Arts in Community Studies, and has also obtained a diploma in Community Leadership Development from the College of the North Atlantic. She currently volunteers as a board member for the People of the DAWN Indigenous Friendship Center in western Newfoundland and contributes to programs and cultural events such as the Sweetgrass Festival. Her various involvements within the community have helped shape her passion for social justice, community development, and cultural awareness.

Ceallaigh S. MacCath-Moran (C.S. MacCath)
Ceallaigh is a PhD candidate in the Folklore Department at Memorial University of Newfoundland, an author, a poet, and a musician. Ceallaigh's research interests include animal rights activism as a public performance of ethical belief, which is the topic of her dissertation, and creative applications of folkloristics for storytellers, which is the topic of her long-running *Folklore & Fiction* podcast. Ceallaigh's ecocritical fairy tale "The Belt and the Necklace" was recently adapted by the Odyssey Theatre in Ottawa as part of *The Other Path* radio play series, and work from her two fiction and poetry collections has been shortlisted for the Washington Science Fiction Association Small Press Award, nominated for the Pushcart Prize, and nominated for the Rhysling Award. Her music is both old and new, inspired by the English and Scottish ballad tradition and rooted in contemporary Paganism.

Elder Albert Marshall
Albert is a leading environmental voice in Unama'ki Cape Breton, Nova Scotia, Canada. He is an advisor to, and is a highly regarded spokesperson for, the Unama'ki Institute of Natural Resources (UINR).

Albert advises and lectures internationally on a wide range of topics including the environment, tribal consciousness and collaboration with non-Indigenous society, traditional healing, traditional teachings, Mi'kmaq orthography and language, and First Nations' vision of science. In collaboration with his late wife Murdena Marshall and Professor Cheryl Bartlett, Albert is the creator of "Etuaptmumk," the "Two-Eyed Seeing" approach which aims to balance Traditional Indigenous Knowledge systems with contemporary science and other cultural perspectives that seek to benefit the earth, humanity, and all living creatures.

Albert works tirelessly to further positive work within Mi'kmaw communities, to seek preservation and understanding of cultural beliefs and practices among all communities, and to affect a strong vision for his people and the future. He considers himself humbly as a conduit transmitting knowledge passed on to him by his ancestors.

Aric McBay
Aric McBay is an organizer, a farmer, and author of seven books, including the climate-fiction novel *Kraken Calling* (2022) and *Full*

Spectrum Resistance (2019), a two-volume guide to building winning movements, both published by Seven Stories Press. He writes and speaks about effective social movements, and has organized many successful campaigns around prisoner justice, Indigenous solidarity, pipelines, unionization, and other causes. Aric works as the Lead, Climate Justice, at the Providence Centre for Peace, Justice, and the Integrity of Creation based in Kingston, Ontario.

David Nibert
David is a scholar/activist and professor of sociology at Wittenberg University. He teaches courses on Animals & Society and Global Injustice. He previously worked as a tenant organizer and as a community activist. He is the author of *Animal Rights/Human Rights: Entanglements of Oppression and Liberation* (Rowman & Littlefield, 2002); *Animal Oppression and Human Violence: Domesecration, Capitalism and Global Conflict* (Columbia University Press, 2013); and he edited *Animal Oppression and Capitalism*, Volumes One and Two (Praeger, 2017). He co-organized the section on Animals and Society of the American Sociological Association.

Catherine O'Brien
Dr Catherine O'Brien has been actively engaged in sustainability efforts locally, nationally, and internationally for more than 30 years. As a participant in the Global Forum of the 1992 Earth Summit, she served as a co-coordinator of the alternative Debt Treaty that was forged by non-governmental organizations from around the world. Her doctoral research at the Barefoot College in Rajasthan, India, explored its pioneering work in education for sustainable community development.

Catherine is a Senior Scholar with the Education Department of Cape Breton University, Canada, where she developed the world's first university course on sustainable happiness based on her pathbreaking concept of sustainable happiness – integrating sustainability principles with positive psychology with the aim of fostering well-being for all, sustainably. Her book on *Education for Sustainable Happiness and Well-being* (2016) is published by Routledge.

Joseph M. Parish
Dr Joe Parish is an Assistant Professor of Anthropology at Cape Breton University. He is a Biological Anthropologist specializing

in Medical Anthropology and Bioarchaeology. His current area of study is human epidemics of the past, present, and future on the Island of Cape Breton (Unama'ki) in Canada's North Atlantic. He is also a professional farmer, owning and operating a small family farm with his wife and their two children. They emphasize organic farming and permaculture methods that meld with the native landscape, encouraging local food diets, biodiversity, and interrelationships with wild animal, fungal, and plant species. He is a descendant of the first Acadian Settlers in Mi'kma'ki and is grateful to be a guest in Unama'ki, the unceded and ancestral territory of the Mi'kmaq.

Sophie Riley
Sophie Riley is an Associate Professor of Law at the University of Technology, Sydney, where she teaches animal law at undergraduate and postgraduate levels. Sophie's research is cross-disciplinary, exploring the legal and ethical implications of wholesale killing as a management tool in environmental regulation. Sophie's research also incorporates strong historical elements, identifying how problems that regulators face today derive from practices of the past, where the environment and animals were seen as resources to be exploited. In the context of farm animals, this research is groundbreaking for its analysis of international veterinary conferences and "quarantine" treaties of the nineteenth and mid twentieth centuries and is published in *The Commodification of Farm Animals* (Springer, 2022).

Margaret Robinson
Dr Margaret Robinson is a bisexual and two-spirit scholar from Eskikewa'kik, Mi'kma'ki, and a member of Lennox Island First Nation. She earned her PhD from the University of Toronto and has been conducting community-based research since 2006. Margaret works as an Associate Professor in the Department of English at Dalhousie University in Nova Scotia, where she holds the Tier II Canada Research Chair in Reconciliation, Gender, and Identity.

Paul Rooke
Paul Rooke is a retired high-school teacher who has traveled extensively throughout the world. During his travels he worked for several months in an Australian slaughterhouse.

Vandana Shiva

Vandana Shiva is Director of the Foundation for Science, Technology & Ecology and Board Member of the International Forum on Globalization. Besides being a physicist, ecologist, activist, editor, and author of numerous books, Vandana Shiva is a tireless defender of the environment. She is the founder of Navdanya, a movement for biodiversity conservation and farmers' rights. She is also the founder and director of the Research Foundation for Science, Technology, and Natural Resource Policy. Shiva fights for changes in the practice and paradigms of agriculture and food.

Intellectual property rights, biodiversity, biotechnology, bioethics, and genetic engineering are among the fields where Shiva has contributed intellectually and through activist campaigns. During the 1970s, she participated in the nonviolent Chipko movement, whose main participants were women. She has assisted grassroots organizations of the Green Movement in Africa, Asia, Latin America, Ireland, Switzerland, and Austria with campaigns against genetic engineering. Shiva has also served as an adviser to governments in India and elsewhere as well as non-governmental organizations, including the International Forum on Globalization, the Women's Environment and Development Organization, the Third World Network, and the Asia Pacific People's Environment Network.

She is the author of over 25 books including *Staying Alive* (North Atlantic Books, 2016), *The Violence of the Green Revolution* (Zed Books, 1991), *Biopiracy: The Plunder of Nature and Knowledge* (North Atlantic Books, 2016), *Monocultures of the Mind* (Zed Books, 1993), *Water Wars: Privatization, Pollution, and Profit* (North Atlantic Books, 2016), and *Stolen Harvest: The Hijacking of the Global Food Supply* (University Press of Kentucky, 2016), as well as over 300 papers in leading scientific and technical journals.

In 1993, she received the Right Livelihood Award, commonly known as the "Alternative Nobel Prize." Other awards include the Order of the Golden Ark, Global 500 Award of the UN, Earth Day International Award, the Lennon Ono Grant for Peace, and the Sydney Peace Prize 2010.

In 2003, *Time* magazine identified Shiva as an "environmental hero," and *Asia Week* has called her one of the five most powerful communicators from Asia.

Av Singh

Av Singh, PhD, Pag, is a proponent of regenerative organic food and medicine production and is engaged in agricultural projects across six continents. Av has had the privilege of visiting over 2,500 farms, which has shaped his extension of holistic, system-based design solutions. Emphasizing a union of traditional knowledge with science, Av works with growers to cultivate an appreciation of plant–soil–microbe interrelationships. Av is a long-time member of the Canadian Organic Growers and the National Farmers' Union, as well as the Vice-President of Régénération Canada. Av is also a faculty member at Earth University (Navdanya) in India, where he delivers courses on agroecology and regenerative organic farming.

Len Vassallo

Len has a Masters degree in marine ecology and over 20 years' experience teaching environmental technology. He has been a subsistence agriculturalist for most of his adult life, but after retiring from an educational career he started a no till biodynamic farm (Blue Heron Farm) to supply locally produced fruit and vegetables. As part of Blue Heron Farm's community outreach program, several community gardens were begun and the local food bank receives weekly deliveries of garden-fresh produce.

Len's work with community organizations has been recognized by Memorial University (outstanding community service award), The Province of Newfoundland and Labrador (environmental lifetime achievement award), The Province of Nova Scotia (recognition in legislature for multi-year donations to food bank) and the City of Corner Brook (Green/Sustainability Award).

References

Action Against Hunger (2023) *World Hunger Facts.* https://www.actionagainsthunger.org/the-hunger-crisis/world-hunger-facts/.

Adams, C. J. (1991) *The Sexual Politics of Meat: A feminist-vegetarian critical theory.* New York: The Continuum Publishing Company.

Adams, C. J. (1994) *Neither Man nor Beast: Feminism and the defense of animals.* New York: The Continuum Publishing Company.

Adams, C. J. (2007) "The war on compassion (2006)." In: Donovan, J. & Adams, C. J. (eds.) *The Feminist Care Tradition in Animal Ethics.* New York: Columbia University Press, pp. 21–34.

Albritton, R. (2009) *Let Them Eat Junk: How capitalism creates hunger and obesity.* London: Pluto Press.

Alfred, T. (2005) *Wasáse: Indigenous pathways of action and freedom.* Peterborough, Ontario: Broadview Press.

Alfred, T. (2009) *Peace, Power, Righteousness: An Indigenous manifesto,* 2nd edn. Toronto, Ontario: Oxford University Press.

Altieri, M. A. (2010) "Agroecology, small farms, and food sovereignty." In: Magdoff, F. & Tokar, B. (eds.) *Agriculture and Food in Crisis: Conflict, resistance, and renewal.* New York: Monthly Review Press, pp. 253–66.

Ameli, K. (2022) *Multispecies Ethnography: Methodology of a holistic research approach of humans, animals, nature, and culture.* Lanham, MD: Lexington Books.

American Society for Microbiology (2011) *FAQ: E. Coli: good, bad, & deadly.* Washington, DC.

Animal Justice (2014) 'Canadian Food Labels Guide', 23 December [Blog]. https://animaljustice.ca/blog/food-labels-guide-non-vegan-ethical-eating-holiday-season.

Arcury, T. A., Mora, D. C. & Quandt, S. A. (2015) "'... you earn money by suffering pain': beliefs about carpal tunnel syndrome among Latino poultry processing workers," *Journal of Immigrant and Minority Health* 17 (3): 791–801.

Arluke, A. (1994) "Managing emotions in an animal shelter." In: Manning,

A. & Serpell, J. (eds.) *Animals and Human Society: Changing perspectives*. London: Routledge, pp. 145–65.

Arluke, A. & Sanders, C. (1996) *Regarding Animals*. Philadelphia, PA: Temple University Press.

ASPCA (n.d.) *Meat, eggs and dairy label guide*. https://www.aspca.org/shop withyourheart/consumer-resources/meat-eggs-and-dairy-label-guide.

ASPCA (2020) *Statement on COVID-19-Related Depopulation of Farm Animals*. https://www.aspca.org/about-us/press-releases/statement-covid-19-related-depopulation-farm-animals.

Azzam, A. (2021) "Is the world converging to a 'Western diet'?," *Public Health Nutrition* 24 (2): 309–17.

Baker, O. (2022) "Some N.S. Mi'kmaw communities still without power after Fiona's devastating winds," *CBC News*, 26 September. https://www.cbc.ca/news/indigenous/fiona-power-outages-eskasoni-membertou-1.6596126.

Balcombe, J. (2006) *Pleasurable Kingdom: Animals and the nature of feeling good*. London: Macmillan.

Balcombe, J. (2010) *Second Nature: The inner lives of animals*. New York: Palgrave Macmillan.

Balcombe, J. (2016) *What a Fish Knows: The inner lives of our underwater cousins*. New York: Scientific American/Farrar, Straus, and Giroux.

Baran, B. E., Rogelberg, S. G., Carello-Lopina, E., Allen, J. A., Spitzmüller, C. & Bergman, M. (2012) "Shouldering a silent burden: the toll of dirty tasks," *Human Relations* 65 (5): 597–626.

Baran, B. E., Rogelberg, S. G. & Clausen, T. (2016) "Routinized killing of animals: Going beyond dirty work and prestige to understand the well-being of slaughterhouse workers," *Organization* 23 (3): 351–69.

Bartlett, C., Marshall, M. & Marshall, A. (2012) "Two-eyed seeing and other lessons learned within a co-learning journey of bringing together Indigenous and mainstream knowledges and ways of knowing," *Journal of Environmental Studies and Sciences* 2 (4): 331–40.

Battiste, M. (2013) *Decolonizing Education: Nourishing the learning spirit*. Vancouver, BC: University of British Columbia Press.

Baur, G. & Kevany, K. M. (2020) "Shifting perceptions through farm sanctuaries." In: Kevany, K. M. (ed.) *Plant-Based Diets for Succulence and Sustainability*. Abingdon, Oxon: Routledge, pp. 123–38.

Baysinger, A. & Kogan, L. R. (2022) "Mental health impact of mass depopulation of swine on veterinarians during COVID-19 infrastructure breakdown," *Frontiers in Veterinary Science* 9: 9842585.

Bekoff, M. (2007) *The Emotional Lives of Animals: A leading scientist explores animal joy, sorrow, and empathy – and why they matter*. Novato, CA: New World Library.

Bekoff, M. (2014) *Rewilding Our Hearts: Building pathways of compassion and coexistence*. Novato, CA: New World Library.

Bekoff, M. (2016) "Stairways to heaven, temples of doom, and humane-washing: A slightly 'better life' for factory farmed animals is not a good life," *Psychology Today*, 17 November.

Belasco, W. J. (2008) *Food: The key concepts*. Oxford: Berg Publishers.

Bell, M. M. & Ashwood, L. (2016) *An Invitation to Environmental Sociology*, 5th edn. Los Angeles, CA: Sage Publications.

Bendell, J. & Read, R. (2021) *Deep Adaptation: Navigating the realities of climate chaos*. Cambridge: Polity Press.

Berreville, O. (2014) "Animal welfare issues in the Canadian dairy industry." In: Sorenson, J. (ed.) *Critical Animal Studies: Thinking the unthinkable*. Toronto, ON: Canadian Scholars Press, pp. 186–207.

Berry, W. (2009a) *Bringing it to the Table: On farming and food*. Berkeley, CA: Counterpoint.

Berry, W. (2009b) *Wendell Berry: the pleasures of eating*. Center for Ecoliteracy. https://www.ecoliteracy.org/article/wendell-berry-pleasures-eating.

Betz, M. V., Nemec, K. B. & Zisman, A. L. (2022) "Plant-based diets in kidney disease: Nephrology professionals' perspective," *Journal of Renal Nutrition* 32 (5): 552–9.

Betz, M. V., Nemec, K. B. & Zisman, A. L. (2023) "Patient perception of plant based diets for kidney disease," *Journal of Renal Nutrition* 33 (2): 243–8.

BIOS (2019) *Ecological Reconstruction*. https://bios.fi/en/we-have-a-plan-ecological-reconstruction/.

Blanchette, A. (2020) *Porkopolis: American animality, standardized life, & the factory farm*. Durham, NC: Duke University Press.

Blanchette, A. (2021) "Ending things, as an end in itself: Notes on quitting American meatpacking," *Anthropology Now* 13 (1): 73–8.

Blas, J. (2020) "Cargill pays record dividend to family owners after profits boom," *Bloomberg News*, 31 July.

Blattner, C. (2020) "From zoonosis to zoopolis," *Derecho Animal: Forum of Animal Law Studies* 11 (4): 41–53.

Blattner, C., Coulter, K., Wadiwel, D. & Kasprzycka, E. (2021) "Covid-19 and capital: Labour studies and nonhuman animals – a roundtable dialogue," *Animal Studies Journal* 10 (1): 240–72.

Bohanec, H. (2013) *The Ultimate Betrayal: Is there happy meat?* Bloomington, IN: iUniverse.

Booker, C. (2020) "Perspective on food-systems" [Keynote Speech, Virtual Food Policy Conference 2020], 28 and 29 July.

Boseley, S. (2003) "Political context of the World Health Organization: Sugar industry threatens to scupper the WHO," *International Journal of Health Services* 33 (4): 831–3.

Brisson, Y. (2014) *Canadian Agriculture at a Glance: The changing face of the Canadian hog industry* [Online], Statistics Canada. https://www150.statcan.gc.ca/n1/en/pub/96-325-x/2014001/article/14027-eng.pdf?st=1fhW1-rI.

References

Broadway, M. J. (2013) "The world 'meats' Canada: Meatpacking's role in the cultural transformation of Brooks, Alberta," *Focus On Geography* 56 (2): 47–53.

Brones, A. (2018) "Karen Washington: It's not a food desert, it's food apartheid," *Guernica: Global Art & Politics*, 7 May.

Brueck, J. F. (2017) "Preface". In: Brueck, J. F. (ed.) *Veganism in an Oppressive World: A vegans-of-color community project*. Sanctuary Publishers, pp. iii–iv.

Buckee, C., Noor, A. & Sattenspiel, L. (2021) "Thinking clearly about social aspects of infectious disease transmission," *Nature* 595 (7866): 205–13.

CalfCare (2019) *Milk feeding*. https://calfcare.ca/management/feeding/milk-feeding/.

Campbell, T. C. & Campbell T. M. (2006) *The China Study: The most comprehensive study of nutrition ever conducted and the startling implications for diet, weight loss and long-term health*, 1st edn. Dallas, TX: BenBella Books.

Campbell, T. C. & Campbell, T. M. (2016) *The China study: The most comprehensive study of nutrition ever conducted and the startling implications for diet, weight loss and long-term health*, rev. & exp. edn. Dallas, TX: BenBella Books.

Canadian Criminal Code, RSC (1985), cC-46, s 445.

Canavan, J. (2017) "'Happy cow', welfarist ideology and the Swedish 'milk crisis': A crisis of romanticized oppression." In: Nibert, D. (ed.) *Animal Oppression and Capitalism: The oppression of nonhuman animals as sources of food*. Santa Barbara, CA: Praeger, pp. 34–55.

Carleton, E. (2022) "Climate change in Africa: What will it mean for agriculture and food security?," International Livestock Research Institute, 28 February.

Carrington, S. (2018) "Food justice and race in the U.S." In: Rodriguez, S. (ed.) *Food Justice: A primer*. Sanctuary Publishers, pp. 175–92.

Carter, B. & Charles, N. (2013) "Animals, agency and resistance," *Journal for the Theory of Social Behaviour* 43 (3): 322–40.

Ceballos, G., Ehrlich, P. R. & Dirzo, R. (2017) "Biological annihilation via the ongoing sixth mass extinction signaled by vertebrate population losses and declines," *Proceedings of the National Academy of Sciences* 114 (30): E6089–96.

Centers for Disease Control and Prevention (2021a) *Chronic Wasting Disease (CWD)*. https://www.cdc.gov/prions/cwd/transmission.html.

Centers for Disease Control and Prevention (2021b) *Prion Diseases*. https://www.cdc.gov/prions/index.html.

Chetty, R., Grusky, D., Hell, M., Hendren, N., Manduca, R. & Narang, J. (2016) "The Fading American Dream: Trends in absolute income mobility since 1940," National Bureau of Economic Research, Working Paper Series no. 22910.

Chomsky, A. (2022) *Is Science Enough?: Forty critical questions about climate justice*. Chicago, IL: Beacon Press.

Chomsky, A., Leech G. & Striffler, S. (2007) *The People Behind Colombia Coal.* Bogotá, Colombia: Casa Editorial Pissando Callos.

Clark, M. A., Domingo, N. G., Colgan, K., et al. (2020) "Global food system emissions could preclude achieving the 1.5° and 2°C climate change targets," *Science* 370 (6517): 705–8.

College of Veterinary Medicine (2016) "Salmonellosis: background, management and control." Animal Health Diagnostic Center. https://www.vet.cornell.edu/animal-health-diagnostic-center/programs/nyschap/modules-documents/salmonellosis-background-management-and-control.

Colling, S. (2021) *Animal Resistance in the Global Capitalist Era.* East Lansing, MI: Michigan State University Press.

Collins, P. H. & Bilge, S. (2020) *Intersectionality*, 2nd edn. Cambridge: Polity Press.

Concepcion, G. P., David, M. P. C. & Padlan, E. A. (2005) "Why don't humans get scrapie from eating sheep? A possible explanation based on secondary structure predictions," *Medical Hypotheses* 64 (5): 919–24.

Cook, C. D. (2004) *Diet for a Dead Planet: How the food industry is killing us.* New York: New Press.

Coulter, K. (2016) *Animals, Work, and the Promise of Interspecies Solidarity.* New York: Palgrave Macmillan.

Coulter, K. (2020) "Towards humane jobs and work-lives for animals." In: Blattner, C. E., Coulter, K. & Kymlicka, W. (eds.) *Animal Labour: A new frontier of interspecies justice?* Oxford: Oxford University Press, pp. 29–47.

Council Regulation (EC) No 1/2005 of 22 December 2004 on the protection of animals during transport and related operations and amending Directives 64/432/EEC and 93/119/EC and Regulation (EC) No 1255/97.

Cowspiracy: The Sustainability Secret (2014) Directed by K. Anderson & K. Kuhn [Documentary]. Los Angeles, CA: A.U.M. Films, First Spark Media & Appian Way Productions.

Cramer, R., Addo, F. R., Campbell, C., et al. (2019) "The emerging millennial wealth gap: Divergent trajectories, weak balance sheets, and implications for social policy," *New America*, 29 October.

Crenshaw, K. (1989) "Demarginalizing the intersection of race and sex: A Black feminist critique of antidiscrimination doctrine, feminist theory and antiracist politics," *University of Chicago Legal Forum* 1989 (1): 139–67.

Croney, C. C. & Anthony, R. (2010) "Engaging science in a climate of values: Tools for animal scientists tasked with addressing ethical problems," *Journal of Animal Science* 88 (13 Suppl.): E75–E81.

Cudworth, E. (2015) "Killing animals: sociology, species relations and institutionalized violence," *The Sociological Review* 63 (1): 1–18.

Dalton, J. (2022) "Shocking farm footage shows piglets with tails cut off and mothers crammed into tiny cages," *The Independent*, 20 October.

Davis, C. G., Dimitri, C., Nehring, R., Collins, L. A., Haley M., Ha, K. & Gillespie, J. (2022) *U.S. Hog Production: Rising output and changing trends in productivity growth*. (ERR-308), US Department of Agriculture, Economic Research Service.

Davis, K. (2014) "Anthropomorphic visions of chickens bred for human consumption." In: Sorenson, J. (ed.) *Critical Animal Studies: Thinking the unthinkable*. Toronto, ON: Canadian Scholars Press, pp. 169–85.

Degrowth Journal (n.d.) *The manifesto of Degrowth Journal*. https://www.degrowthjournal.org/the-manifesto-of-degrowth-journal/.

de Jong, M. D. T. & Huluba, G. (2020) "Different shades of greenwashing: Consumers' reactions to environmental lies, half-lies, and organizations taking credit for following legal obligations," *Journal of Business and Technical Communication* 34 (1): 38–76.

Di Concetto, A., Duval, E. & Lecorps, B. (2022) "Animal welfare standards in EU organic certification," The European Institute for Animal Law & Policy. https://animallaweurope.com/wp-content/uploads/2022/07/Research-Note-5-Animal-Welfare-Standards-in-the-EU-Organic-Certification-1.pdf.

Dillard, C. & Blome, J. (2022) "A step toward defining constitutional freedom as including rights to a restored biodiverse environment and equal opportunities in life," *Trends: ABA Section of Environment, Energy, and Resources Newsletter*. 53 (6), 15–17. https://www.americanbar.org/groups/environment_energy_resources/publications/trends/2021-2022/july-aug-2022/a-step-toward-defining-constitutional-freedom/.

Dillard, J. (2008) "A slaughterhouse nightmare: Psychological harm suffered by slaughterhouse employees and the possibility of redress through legal reform," *Georgetown Journal on Poverty Law & Policy* 15 (2): 391–408.

Donaldson, S. & Kymlicka, W. (2020) "Animal labour in a post-work society." In: Blattner, C. E., Coulter, K. & Kymlicka, W. (eds.) *Animal Labour: A new frontier of interspecies justice?* Oxford: Oxford University Press, pp. 207–28.

Donovan, J. & Adams, C. J. (2007) "Introduction." In: Donovan, J. & Adams, C. J. (eds.) *The Feminist Care Tradition in Animal Ethics: A Reader*. New York: Columbia University Press, pp. 1–15.

Dryden, J. & Rieger, S. (2020) "Inside the slaughterhouse: North America's largest single coronavirus outbreak started at this Alberta meat-packing plant – take a look within," *CBC News*, 6 May.

Dryden, J. & Rieger, S. (2021) "The human machine: A COVID-19 outbreak at an Alberta slaughterhouse claimed 3 lives. A year later, at another plant, it happened again," *CBC News*, 4 May.

Dufour, S., Labrie, J. & Jacquesa, M. (2019) "The mastitis pathogens culture collection," *Microbiology Resource Announcements* 8 (15).

Duverger, T. (2023) "Degrowth: the history of an idea," *Digital Encyclopedia of European History*, 22 June.

Economist Intelligence (2022) *Democracy Index 2022*. https://www.eiu.com/n/campaigns/democracy-index-2022/.

Einstein, A. (2015) Quoted in *The Liberator Magazine*, 21 December.

Ellis, C. (2014) "Boundary labor and the production of emotionless commodities: The case of beef production," *The Sociological Quarterly* 55 (1): 92–118.

Engebretson, M. (2008) "North America." In: Appleby, M. C. (ed.) *Long Distance Transport and Welfare of Farm Animals*. Oxford: CABI, pp. 218–60.

FAO, IFAD, UNICEF, WFP & WHO (2018) *The State of Food Security and Nutrition in the World: Building climate resilience for food security and nutrition*. Rome: FAO.

FAO, IFAD, UNICEF, WFP & WHO (2022) *The State of Food Security and Nutrition in the World (SOFI) Report: Repurposing food and agricultural policies to make healthy diets more affordable*. Rome: FAO.

FAO, OIE & WHO (2010) *Influenza and Other Emerging Zoonotic Diseases at the Human-Animal Interface: FAO/OIE/WHO joint scientific consultation 27–29 April 2010*. Rome: FAO.

Farinosi, F., Giupponi, C., Reynaud, A., et al. (2018) "An innovative approach to the assessment of hydro-political risk: A spatially explicit, data driven indicator of hydro-political issues," *Global Environmental Change* 52: 286–313.

Farm System Reform Act (2019) S.3221. 116th Congress Senate Bill.

Fischer, F. (2019) "Knowledge politics and post-truth in climate denial: On the social construction of alternative facts," *Critical Policy Studies* 13 (2): 133–52.

Fish Count (n.d.) *Fish count estimates*. http://fishcount.org.uk/fish-count-estimates-2.

Fitzgerald, A. J. (2010) "A social history of the slaughterhouse: From inception to contemporary implications," *Society for Human Ecology* 17 (1): 58–69.

Fitzgerald, A. J. (2015) *Animals as Food: (Re)connecting production, processing, consumption, and impacts*. East Landing, MI: Michigan State University Press.

Fitzgerald, A. J. (2019) *Animal Advocacy and Environmentalism*. Cambridge: Polity Press.

Fitzgerald, A., Kalof, L. & Dietz, T. (2009) "Slaughterhouses and increased crime rates: An empirical analysis of the spillover from 'the jungle' into the surrounding community," *Organization & Environment* 22 (2): 158–84.

Foer, J. S. (2009) *Eating Animals*. New York: Little, Brown and Company.

Food and Agriculture Organization of the United Nations (2006) *Policy Brief: Food security – Issue 2*. Rome: FAO.

Food and Agriculture Organization of the United Nations (2013) *Food Wastage Footprint: Impacts on natural resources – summary report*. Rome: FAO. https://www.fao.org/3/i3347e/i3347e.pdf.

Food and Agriculture Organization of the United Nations (2019) *State of Food and Agriculture: Moving forward on food loss and waste reduction.* Rome: FAO. https://www.fao.org/3/ca6030en/ca6030en.pdf.

Food and Agriculture Organization of the United Nations (2022) *The State of World Fisheries and Aquaculture: Towards blue transformation.* Rome: FAO.

Food and Agriculture Organization of the United Nations (2023) *Fishing Safety.* Rome: FAO. https://www.fao.org/fishing-safety/en/.

Food and Agriculture Organization of the United Nations (n.d.) *Agroecology Knowledge Hub.* https://www.fao.org/agroecology/home/en/.

Food and Agriculture Organization of the United Nations & World Food Programme (2022) *Hunger Hotspots – FAO-WFP early warnings on acute food insecurity: October 2022 to January 2023 outlook.* Rome: FAO; WFP.

Food Inc. (2008) Directed by Robert Kenner [Documentary]. Toronto, Ontario: International Film Festival.

Forrest, S. (2022) "What is driving the high suicide rate among farmers?," *Illinois News Bureau*, 12 December.

Foster, J. B. & Burkett, P. (2017) *Marx and the Earth: An anti-critique.* Chicago, IL: Haymarket Books.

Frank, L., Fisher, L. & Saulnier, C. (2021) *Report Card on Child and Family Poverty in Nova Scotia: Worst provincial performance over 30 years.* Nova Scotia: Canadian Centre for Policy Alternatives. https://policyalternatives.ca/sites/default/files/uploads/publications/Nova%20Scotia%20Office/2021/11/2021reportcardonchildandfamilypovertyinNovaScotia.pdf

Freeman, A. (2021) "COVID and meat-packing: What Canada's co-op movement doesn't want you to know," *IPolitics*, 4 March.

Galtung, J. (1969) "Violence, peace, and peace research," *Journal of Peace Research* 6 (3): 167–91.

Gibbs, T. (2017) *Why the Dalai Lama is a Socialist: Buddhism, socialism and the compassionate society.* London: Zed Books.

Gibbs, T. & Harris, T. (2020) "The vegan challenge is a democracy issue: Citizenship and the living world." In: Kevany, K. M. (ed.) *Plant-Based Diets for Succulence and Sustainability.* Abingdon, Oxon: Routledge, pp. 139–54.

Gillespie, K. (2011) "How happy is your meat? Confronting (dis)connectedness in the 'alternative' meat industry," *The Brock Review* 12 (1): 100–28.

Gillespie, K. (2014) "Sexualized violence and the gendered commodification of the animal body in Pacific Northwest US dairy production," *Gender, Place & Culture* 21 (10): 1321–37.

Gillespie, K. (2016) "Witnessing animal others: Bearing witness, grief, and the political function of emotion," *Hypatia* 31 (3): 572–88.

Gillespie, K. (2018) *The Cow with Ear Tag #1389.* Chicago, IL: University of Chicago Press.

Gillespie, K. & Lopez, P. J. (2019) *Vulnerable Witness: The politics of grief in the field.* Oakland, CA: University of California Press.

Gillespie, K. & Lopez, P. (2020) *Economies of Death: Economic logics of killable life and grievable death*. London: Routledge.

Goodland, R., & Anhang, J. (2009) "Livestock and climate change: What if the key actors are ... cows, pigs, and chickens?," *World Watch Institute* 22 (6): 10–19.

Government of Canada (2016) *Causes of salmonellosis*. https://www.canada.ca/en/public-health/services/diseases/salmonellosis-salmonella/causes.html.

Government of Canada (2021a) *Food safety for people with a weakened immune system*. https://www.canada.ca/en/health-canada/services/food-safety-vulnerable-populations/food-safety-people-with-weakened-immune-system.html.

Government of Canada (2021b) *Guidance on essential services and functions in Canada during the COVID-19 pandemic*. https://www.publicsafety.gc.ca/cnt/ntnl-scrt/crtcl-nfrstrctr/esf-sfe-en.aspx.

Government of Canada (2022) *Fact sheet – Classical bovine spongiform encephalopathy (classical BSE)*. https://inspection.canada.ca/animal-health/terrestrial-animals/diseases/reportable/bovine-spongiform-encephalopathy/classical-bse/eng/1650551381899/1650551382212.

Government of Canada, Canada Food Inspection Agency (2022) *Livestock and poultry transport in Canada*. https://inspection.canada.ca/DAM/DAM-animals-animaux/STAGING//text-texte/livestock_transport_pdf_1528296360187_eng.pdf.

Graf, A. (2021) "The climate crisis is a privilege crisis," *Youth4Nature*, 24 September.

Grandin, T. (2018) *Proper cattle restraint for stunning*. https://www.grandin.com/humane/restrain.slaughter.html.

Gray, A. (2016) "Udder justice: The dairy cow's experience of milk production regulations in Canada," *Contemporary Justice Review* 19 (2): 221–9.

Greenebaum, J. B. (2017) "Questioning the concept of vegan privilege: A commentary," *Humanity & Society* 41 (3): 355–72.

Greger, M. (2007) "The human/animal interface: Emergence and resurgence of zoonotic infectious diseases," *Critical Reviews in Microbiology* 33 (4): 243–99.

Greger, M. (2020) *How to Survive a Pandemic*. New York: Flatiron Books.

Greger, M. (2021) "Primary pandemic prevention," *American Journal of Lifestyle Medical* 15 (5): 498–505.

Griffith-Greene, M. (2014) "Baby chickens 'cooked alive' at hatchery, animal rights group contends," *CBC News*, 15 April.

Gruen, L. (2015) *Entangled Empathy: An alternative ethic for our relationship with animals*. Brooklyn, NY: Lantern Books.

Guillemette, A. R. & Cranfield, J. A. L. (2012) "Food expenditures: the effect of a vegetarian diet and organic foods," University of Guelph. https://

www.researchgate.net/publication/254385228_Food_Expenditures_The _Effect_of_a_Vegetarian_Diet_and_Organic_Foods.

Guthman, J., Broad, G., Klein, K. & Landecker, H. (2014) "Beyond the sovereign body," *Gastronomica: The Journal of Food and Culture* 14 (3) 46–55.

Haedicke, M. A. (2013) "From collective bargaining to social justice certification: Workers' rights in the American meatpacking industry," *Sociological Focus* 46 (2): 119–37.

Haider, A. & Roque, L. (2021) "New poverty and food insecurity data illustrate persistent racial inequities," *Center for American Progress*, 29 September.

Hamilton, L. & McCabe, D. (2016) "It's just a job: Understanding emotion work, de-animalization and the compartmentalization of organized animal slaughter," *Organization* 23 (3): 330–50.

Hannan, J. (ed.) (2020) *Meatsplaining: The animal agriculture industry and the rhetoric of denial*. Sydney: Sydney University Press.

Harfeld, J. L., Cornou, C., Kornum, A. & Gjerris, M. (2016) "Seeing the animal: On the ethical implications of de-animalization in intensive animal production systems," *Journal of Agricultural & Environmental Ethics* 29 (3): 407–23.

Harper, A. B. (2020) *Sistah Vegan: Black women speak on food, identity, health, and society*, 10th anniversary edn. Brooklyn, NY: Lantern Publishing and Media.

Harper, C. L. & Le Beau, B. F. (2002) *Food, Society, and Environment*. Upper Saddle River, NJ: Prentice Hall.

Harris, T. (2017) "The problem is not the people, it's the system: The Canadian animal industrial complex." In: D. Nibert (ed.) *Animal Oppression and Capitalism: The oppression of nonhuman animals as sources of food*. Santa Barbara, CA: Praeger, pp. 57–75.

Harris, T. (2018) *The Tiny House Movement: Challenging our consumer culture*. Lanham, MD: Lexington Books.

Harrison, R. (1964) *Animal Machines: The new factory farming industry*. London: Vincent Stuart Publishers.

Harwatt, H. (2019) "Including animal to plant protein shifts in climate change mitigation policy: A proposed three-step strategy," *Climate Policy* 19 (5): 533–41.

Head, L. (2016) *Hope and Grief in the Anthropocene: Re-conceptualising human-nature relations*. Abingdon, Oxon: Routledge.

Health of Animals Act (2019) "Regulations amending the health of animals regulations: SOR/2019-38," *Government of Canada; Canada Gazette* 153 (4).

Health Canada (2019) *Canada's Dietary Guidelines for Health Professionals and Policy Makers*. Government of Canada. https://food-guide.canada.ca/sites /default/files/artifact-pdf/CDG-EN-2018.pdf.

Held, D. (1995) *Democracy and Global Order: From the modern state to cosmopolitan governance.* Stanford, CA: Stanford University Press.

Heredia, N. & García, S. (2018) "Animals as sources of food-borne pathogens: A review," *Animal Nutrition* 4 (3): 250–5.

Herman, E. S. & Chomsky, N. (2002) *Manufacturing Consent: The political economy of the mass media*, 2nd edn. New York: Pantheon Books.

Herring, D. A. & Swedlund, A. C. (2010) "Plagues and epidemics in anthropological perspective." In: Herring, D. A. & Swedlund, A. C. (eds.) *Plagues and Epidemics: Infected spaces past and present.* Oxford: Berg Publishers, pp. 1–20.

Hoekstra, A. Y. (2012) "The hidden water resource use behind meat and dairy," *Animal Frontiers* 2 (2): 3–8.

Holt-Giménez, E. (2010) "From food crisis to food sovereignty: The challenge of social movements." In: Magdoff, F. & Tokar, B. (eds.) *Agriculture and Food in Crisis: Conflict, resistance, and renewal.* New York: Monthly Review Press, pp. 207–23.

Holt-Giménez, E. (2017a) *A Foodie's Guide to Capitalism: Understanding the political economy of what we eat.* New York: Monthly Review Press.

Holt-Giménez, E. (2017b) "Introduction: Agrarian questions and the struggle for land justice in the United States." In: Williams, J. M. & Holt-Giménez, E. (eds.) *Land Justice: Re-imagining land, food, and the commons in the United States.* Oakland, CA: Food First Books, pp. 1–14.

Human Rights Watch (2020) *Colombia: Indigenous kids at risk of malnutrition, death – improve Wayuu's access to food, water, health during Covid-19.* https://www.hrw.org/news/2020/08/13/colombia-indigenous-kids-risk-malnutrition-death.

Human Rights Watch (2022) "Colombia." In: Human Rights Watch (ed.) *World Report: Events of 2021.* New York: Human Rights Watch, pp. 179–89.

Humane Canada (n.d.) *Realities of farming in Canada.* https://humanecanada.ca/our-work/focus-areas/farmed-animals/realities-of-farming-in-canada/.

Hume, S. (2004). "Fishing for Answers." In: Hume, S., Morton, A., Keller, B. C., Leslie, R. M., Langer, O. & Staniford, D. (eds.) *A Stain Upon the Sea: West coast salmon farming.* Madeira Park, BC: Harbour Publishing, pp. 17–77.

Hussein, O. H., Abdel-Hameed, K. G. & El-Malt, L. M. (2022) "Prevalence and public health hazards of subclinical mastitis in dairy cows," *SVU-International Journal of Veterinary Sciences* 5 (3): 52–64.

Inoue, T. (2017) "Oceans filled with agony: Fish oppression driven by capitalist commodification." In: Nibert, D. (ed.) *Animal Oppression and Capitalism: The oppression of nonhuman animals as sources of food.* Santa Barbara, CA: Praeger, pp. 96–117.

Intergovernmental Science-Policy Platform on Biodiversity and Ecosystem

Services (2019) *Media Release: Nature's dangerous decline "unprecedented"; species extinction rates "accelerating,"* 5 May.

Jacques, J. R. (2015) "The slaughterhouse, social disorganization, and violent crime in rural communities," *Society & Animals* 23 (6): 594–612.

Jala, D (2022) "Neighbours helping neighbours: Hundreds visit impromptu Cape Breton kitchen," *Saltwire*, 26 September.

Jamail, D. (2020) *The End of Ice: Bearing witness and finding meaning in the path of climate disruption.* New York: The New Press.

Jones, A. (2022) "Youth climate activists to challenge Ontario government in court over greenhouse gas targets," *CBC News*, 11 September.

Joy, M. (2010) *Why We Love Dogs, Eat Pigs and Wear Cows: An introduction to carnism.* San Francisco, CA: Conari Press.

Jung, Y., Jang, H. & Matthews, K. (2014) "Effect of the food production chain from farm practices to vegetable processing on outbreak incidence," *Microbial Biotechnology* 7 (6): 517–27.

Kabir, S. M. L. (2010) "Avian colibacillosis and salmonellosis: A closer look at epidemiology, pathogenesis, diagnosis, control and public health concerns," *International Journal of Environmental Research and Public Health* 7 (1): 89–114.

Kateman, B. (2022) *Meat Me Halfway: How changing the way we eat can improve our lives and save the planet.* Lanham, MD: Rowman & Littlefield.

Kevany, K. (2022) "Let's make plant-based foods the default on campus," *University Affairs*, 13 October.

Kevany, S. (2020a) "Alleged animal abuse in US dairy sector under investigation: Claims of violent treatment and cows being passed off as organic have been presented to the Department of Agriculture," *The Guardian*, 15 October.

Kevany, S (2020b) "Millions of US farm animals to be culled by suffocation, drowning and shooting," *The Guardian*, 19 May.

Kimmerer, R. W. (2013) *Braiding Sweetgrass: Indigenous wisdom, scientific knowledge and the teachings of plants.* Minneapolis, MN: Milkweed Editions.

Kimmerer, R. W. (2021) *The Democracy of Species.* London: Penguin.

King, B. J. (2021) *Animals' best friends: Putting compassion to work for animals in captivity and in the wild.* Chicago, IL: University of Chicago Press.

Klein, N. (2008) *The Shock Doctrine: The rise of disaster capitalism.* Toronto, Ontario: Vintage Canada.

Klein, N. (2014) *This Changes Everything: Capitalism vs. the climate.* Toronto, Ontario: Vintage Canada.

Klinenberg, E. (2018) *Palaces for the People: How social infrastructure can help fight inequality, polarization, and the decline of civic life.* New York: Crown.

Ko, A. & Ko, S. (2020) *Aphro-ism: essays on pop culture, feminism, and Black veganism from two sisters.* Brooklyn, NY: Lantern Publishing and Media.

Koneswaran, G. & Nierenberg, D. (2008) "Global farm animal production and global warming: Impacting and mitigating climate change," *Environmental Health Perspectives* 116 (5): 578–82.

Kopecky, A. (2022) "Exposing animal abuse may land Canadian farm activists in jail," *Mother Jones*, 13 October.

Kovach, M. (2009) *Indigenous Methodologies: Characteristics, conversations and contexts*. Toronto, ON: University of Toronto Press.

Kumar, K. & Makarova, E. (2008) "The portable home: The domestication of public space," *Sociological Theory* 26 (4): 324–43.

Kurt, T. D. & Sigurdson, C. J. (2016) "Cross-species transmission of CWD prions," *Prion* 10 (1): 83–91.

Kyeremateng-Amoah, E., Nowell, J., Lutty, A., Lees, P. S. J. & Silbergeld, E. K. (2014) "Laceration injuries and infections among workers in the poultry processing and pork meatpacking industries: Injuries in meat and poultry workers," *American Journal of Industrial Medicine* 57 (6), 669–82.

Labchuk, C (2020) "Brutality of the meat industry is on display during COVID-19 pandemic," *Toronto Star*, 21 May.

LaBronx, H. (2018) *A pro-intersectional approach in the fight for animal rights* [Keynote speech, Veganes Sommerfest Berlin], 24–25 August.

Ladd, A. E. & Edward, B. (2002) "Corporate swine and capitalist pigs: A decade of environmental injustice and protest in North Carolina," *Social Justice* 29 (3): 26–46.

LaFortune, R (2020) "'My Fear is Losing Everything': The climate crisis and First Nations' right to food in Canada", Human Rights Watch Report, October 21. https://www.hrw.org/report/2020/10/21/my-fear-losing-everything/climate-crisis-and-first-nations-right-food-canada.

LearnRidge (2023) *Mi'kmaq culture and resilience*. https://nscs.learnridge.com/topic/mi-kmaq-culture-and-resilience/.

Lee, H. & Yoon, Y. (2021) "Etiological agents implicated in foodborne illness world wide," *Food Science of Animal Resources* 41 (1): 1–7.

Leech, G. (2011) *The Farc: The longest insurgency*. London: Zed Books.

Leech, G. (2012) *Capitalism: A structural genocide*. London: Zed Books.

Leopold, A. (1949) *A Sand County Almanac*. Oxford: Oxford University Press.

Li, Z., Chen, Y. & Zhan, Y. (2022) "Building community-centered social infrastructure: A feminist inquiry into China's COVID-19 experiences," *Economia Politica* 39 (1): 303–21.

Lipscomb, H. J., Dement, J. M., Epling, C. A., Gaynes, B. N., McDonald, M. A. & Schoenfisch, A. L. (2007) "Depressive symptoms among working women in rural North Carolina: A comparison of women in poultry processing and other low-wage jobs," *International Journal of Law and Psychiatry*, 30 (4): 284–98.

López Zuleta, D. (2015) "The case of the 5000 children who have died of hunger in La Guajira reaches the OAS," *Las2Orillas*, 16 March.

Lorenzen, J. A. (2014) "Green consumption and social change: Debates over responsibility, private action, and access," *Sociology Compass* 8 (8): 1063–81.

Louv, R. (2012) *The Nature Principle: Reconnecting with life in a virtual age.* Chapel Hill, NC: Algonquin Books.

Lubin, R. (2019) "Sharks' fins are cut while still alive in sick trade that could wipe species out," *The Mirror*, 19 March.

Lund, V., Mejdell, C., Röcklinsberg, H., Anthony, R. & Håstein, T. (2007) "Expanding the moral circle: Farmed fish as objects of moral concern," *Diseases of Aquatic Organisms* 75 (2): 109–18.

Lusk, J. L. & Norwood, F. B. (2016) "Some vegetarians spend less money on food, others don't," *Ecological Economics* 130: 232–42.

Lymbery, P. (2020) "Covid-19: How industrial animal agriculture fuels pandemics," *Derecho Animal* 11 (4): 141–9.

Lynch, J., Cain, M., Frame, D. & Pierrehumbert, R. (2021) "Agriculture's contribution to climate change and role in mitigation is distinct from predominantly fossil CO_2-emittings sectors," *Frontiers in Sustainable Food Systems* 4: 518039.

Maathai, W. (2009) *The Challenge for Africa.* New York: Pantheon Books.

Maathai, W. (2010) *Replenishing the Earth: Spiritual values for healing ourselves and the world.* New York: Random House, Doubleday Religion.

Mackenzie, J. S. & Jeggo, M. (2013) "Reservoirs and vectors of emerging viruses," *Current Opinion in Virology* 3 (2): 170–9.

Marshall, A. (2017) *Climate change, drawdown & the human prospect: A retreat for empowering our climate future for rural communities* [Talk given at Thinkers Lodge, Pugwash, Nova Scotia], 28 September–1 October.

Marshall, A. (2022) *Reconciliation with the Earth* [Presentation to Allison Bernard Memorial High School, Eskasoni, Nova Scotia], 10 January.

Martin, A. (n.d.) "Seven ways to fight for food justice," *Food Tank*. https://foodtank.com/news/2021/02/ways-to-fight-for-food-justice/.

Martin, J., Maris, V. & Simberloff, D. S. (2016) "The need to respect nature and its limits challenges society and conservation science," *Proceedings of the National Academy of Sciences* 113 (22): 6105–12.

Martínez-González, M. A., Sánchez-Tainta, A., Corella, D., et al. (2014) "Provegetarian food pattern and reduction in total mortality in the Prevención con Dieta Mediterránea (PREDIMED) study," *The American Journal of Clinical Nutrition* 100 (1): 320S–328S.

Marty, C. (2023) "André Gorz's Vision for Autonomy and Radical Frugality," *Green European Journal*, 9 February.

Mason, J. (1998) *An Unnatural Order: Why we are destroying the planet and each other.* New York: Continuum.

Masson, J. M. (2004) *The Pig Who Sang to the Moon: The emotional world of farm animals.* New York: Ballantine Books.

Masson, J. M. (2009) *The Face on Your Plate: The truth about food.* New York: W. W. Norton.
Maurer, J. M., Schaefer, J. M., Rupper, S. & Corley, A. (2019) Acceleration of ice loss across the Himalayas over the past 40 years. *Science Advances* 5 (6): eaav7266.
Maxwell, J. A. (ed.) (2005) *Qualitative Research Design: An interactive approach*, 2nd edn. Thousand Oaks, CA: Sage.
McGranahan, C. (2016) "Theorizing refusal: An introduction," *Cultural Anthropology* 31 (3): 319–25.
McLoughlin, E. (2019) "Knowing cows: Transformative mobilizations of human and non-human bodies in an emotionography of the slaughterhouse," *Gender, Work & Organization* 26 (3): 322–42.
Medora, D. (2014) "Spinning the pig: The language of industrial pork production." In: Sorenson, J. (ed.) *Critical Animal Studies: Thinking the unthinkable.* Toronto, ON: Canadian Scholars Press, pp. 208–15.
Miller, C. D. M. & Rudolphi, J. M. (2022) "Characteristics of suicide among farmers and ranchers: Using the CDC NVDRS 2003–2018," *American Journal of Industrial Medicine* 65 (8): 675–89.
Minto, R. (2022) "Americans can now expect to live three years less than Cubans," *Newsweek*, 9 February.
Montenegro de Wit, M. (2021) "What grows from a pandemic? Toward an abolitionist agroecology," *The Journal of Peasant Studies* 48 (1): 99–136.
Mora, D. C., Arcury, T. A. & Quandt, S. A. (2016) "Good job, bad job: Occupational perceptions among Latino poultry workers," *American Journal of Industrial Medicine* 59 (10): 877–86.
Mosby, I. & Rotz, S. (2020) "As meat plants shut down, COVID-19 reveals the extreme concentration in our food supply," *The Globe and Mail*, 29 April.
National Farm Animal Care Council (2014) *Code of Practice for the Care and Handling of Pigs.* https://www.nfacc.ca/pdfs/codes/pig_code_of_practice.pdf.
National Farm Animal Care Council (2017) *Code of Practice for the Care and Handling of Pullets and Laying Hens.* https://www.nfacc.ca/pdfs/codes/Pullets%20and%20laying%20hens%20Code_HARrev_21_FINAL.pdf.
National Farm Animal Care Council (2023) *Code of Practice for the Care and Handling of Farm Animals.* https://www.nfacc.ca/codes-of-practice.
National Wildlife Federation (NWF) (2023) *Ecosystem Services.* https://www.nwf.org/Educational-Resources/Wildlife-Guide/Understanding-Conservation/Ecosystem-Services.
National Wildlife Health Center (2022) "Distribution of highly pathogenic avian influenza in North America, 2021/2022," US Geographical Survey, 27 November. https://www.usgs.gov/centers/nwhc/science/distribution-highly-pathogenic-avian-influenza-north-america-20212022.

NASA (2023) *The Effects of Climate Change*. Global Climate Change: Vital Signs of the Planet. https://climate.nasa.gov/effects/.

Navdanya International (2023) https://navdanyainternational.org/.

Naylor. G. (2017) "Agricultural parity for land de-commodification." In: Williams, J. M. & Holt-Gimenez, E. (eds.) *Land Justice: Re-imagining land, food, and the commons on the United States*. Oakland, CA: Food First Books/Institute for Food and Development Policy, pp. xviii–xxii.

Nero's Guests (2009) Directed by D. Bhatia & P. Sainath [Documentary]. London, New York.

Nibert, D. (2002) *Animal Rights Human Rights: Entanglements of oppression and liberation*. Lanham, MD: Rowman & Littlefield.

Nibert, D. (2013) *Animal Oppression & Human Violence: Domesecration, capitalism, and global conflict*. New York: Columbia University Press.

Nibert, D. (2014) "Animals, immigrants, and profits: Slaughterhouses and the political economy of oppression." In: Sorenson, J. (ed.) *Critical Animal Studies: Thinking the unthinkable*. Toronto, ON: Canadian Scholars Press, pp. 3–17.

Nibert, D. (2017) "Introduction." In: Nibert, D. (ed.) *Animal Oppression and Capitalism: The oppression of nonhuman animals as sources of food*. Santa Barbara, CA: Praeger, pp. xi–xxiv.

Nixon, R. (2011) *Slow Violence and the Environmentalism of the Poor*. Cambridge, MA: Harvard University Press.

Norgaard, K. M. (2019) "Making sense of the spectrum of climate denial," *Critical Policy Studies* 13 (4): 437–41.

Norwood, F. B. & Lusk, J. L. (2011) *Compassion by the Pound: The economics of farm animal welfare*. Oxford: Oxford University Press.

Noske, B. (1997) *Beyond Boundaries: Human and Animals*. Montreal, Quebec: Black Rose Books.

O'Brien, C. (2016) *Education for Sustainable Happiness and Well-Being*. New York: Routledge.

Okorie-Kanu, Ezenduka, E. V., Okorie-Kanu, C. O., Ugwu, L. C. & Nnamani, U. J. (2016) "Occurrence and antimicrobial resistance of pathogenic *Escherichia coli* and *Salmonella* spp. in retail raw table eggs sold for human consumption in Enugu state, Nigeria," *Veterinary World* 9 (11): 1312–19.

Olsson, L., Barbosa, H., Bhadwal, S., et al. (2022) "Land degradation." In: Shukla, P. R., Skea, J., Calvo Buendia, E., et al. (eds.) *Climate Change and Land: An IPCC special report on climate change, desertification, land degradation, sustainable land management, food security, and greenhouse gas fluxes in terrestrial ecosystems*. Cambridge: Cambridge University Press, pp. 345–436.

Oster, R. T., Grier, A., Lightning, R., Mayan, M. J. & Toth, E. L. (2014) "Cultural continuity, traditional Indigenous language, and diabetes in

Alberta First Nations: A mixed methods study," *International Journal for Equity in Health* 13 (1): 92.

Oxfam International (2022) "'Terrifying prospect' of over a quarter of a billion more people crashing into extreme levels of poverty and suffering this year," 12 April. https://www.oxfam.org/en/press-releases/terrifying-prospect-over-quarter-billion-more-people-crashing-extreme-levels-poverty.

Pachirat, T. (2011) *Every Twelve Seconds: Industrialized slaughter and the politics of sight.* New Haven, CT: Yale University Press.

Patel, R. (2009) *Stuffed and Starved: The hidden battle for the world's food system.* Toronto, ON: Harper Perennial.

Patel, R. & Moore, J. (2017) *A History of the World in Seven Cheap Things: A guide to capitalism, nature, and the future of the planet.* Berkeley, CA: University of California Press.

Peden, A. H., Suleiman, S. & Barria, M. A. (2021) "Understanding intra-species and inter-species prion conversion and zoonotic potential using protein misfolding cyclic amplification," *Frontiers in Aging Neuroscience* 13: 716452.

Pempek, J. A., Schuenemann, G. M., Holder, E. & Habing, G. G. (2017) "Dairy calf management – A comparison of practices and producer attitudes among conventional and organic herds," *Journal of Dairy Science* 100 (10): 8310–21.

Penrod, G. (2004) "Letting loose the images of war," *Reporters Committee for Freedom of the Press.* https://www.rcfp.org/journals/the-news-media-and-the-law-summer-2004/letting-loose-images-war/.

Pew Commission on Industrial Farm Animal Production (2008) *Putting Meat on the Table: Industrial farm animal production in America.* https://www.pewtrusts.org/en/research-and-analysis/reports/0001/01/01/putting-meat-on-the-table.

Piketty, T. (2013) *Capital in the Twenty-First Century* (trans A. Goldhammer). Cambridge, MA: The Belknap Press of Harvard University Press.

Polansek, T. & Huffstutter, P. J. (2020) "Piglets aborted, chickens gassed as pandemic slams meat sector," *Reuters*, 27 April.

Political Database of the Americas (2008) *Constitution of the Republic of Ecuador.* https://pdba.georgetown.edu/Constitutions/Ecuador/english08.html.

Pollan, M. (2006) *The Omnivore's Dilemma: A natural history of four meals.* New York: Penguin Press.

Pollan, M. (2008) *In Defense of Food: An eater's manifesto.* New York: Penguin Press.

Poore, J. & Nemecek, T. (2018) "Reducing food's environmental impacts through producers and consumers," *Science* 360 (6392): 987–92.

Porcher, J. (2011) "Relationship between workers and animals in the pork

industry: A shared suffering," *Journal of Agricultural & Environmental Ethics* 24 (1): 3–17.
PROOF (2022) *Who are most at risk of household food insecurity?* https://proof.utoronto.ca/food-insecurity/who-are-most-at-risk-of-household-food-insecurity/.
Quammen, D. (2012) *Spillover: Animal infections and the next human pandemic*. New York: W.W. Norton.
Reinhardt, T. A., Lippolis, J. D., McCluskey, B. J., Goff, J. P. & Horst, R. (2011) "Prevalence of subclinical hypocalcemia in dairy herds," *The Veterinary Journal* 188 (1): 122–4.
Reyes-Illg, G., Martin, J. E., Mani, I., Reynolds, J. & Kipperman, B. (2022) "The rise of heatstroke as a method of depopulating pigs and poultry: Implications for the US veterinary profession," *Animals* 13 (1): 140.
Riley, S. (2019) "Listening to nature's voice: Invasive species, earth jurisprudence and compassionate conservation," *Asia Pacific Journal of Environmental Law* 22 (1): 117–36.
Riley, S. (2022) *The Commodification of Farm Animals*, vol. 21. Sydney, Australia: Springer.
Ritchie, H. (2021) "Smallholders produce one-third of the world's food, less than half of what many headlines claim," *Our World in Data*. https://ourworldindata.org/smallholder-food-production.
Ritchie, H., Rosado, P. & Roser, M. (2017a; 2019a) "Meat and dairy production," *Our World in Data*. https://ourworldindata.org/meat-production.
Ritchie, H., Rosado, P. & Roser, M. (2017b; 2019b) "Livestock counts." In: Ritchie, H., Rosado, P. & Roser, M. (eds.) *Meat and Dairy Production*. https://ourworldindata.org/meat-production.
Ritchie, H. & Roser, M. (2021) "Soy." In: Ritchie, H. & Roser, M. (eds.) *Forests and Deforestation*. Our World in Data. https://ourworldindata.org/soy.
Ritzer, G. (2015) *The McDonaldization of Society*, 8th edn. Los Angeles, CA: Sage.
Robinson, M. (2013) "Veganism and Mi'kmaq legends," *Canadian Journal of Native Studies* 33 (1): 189–96.
Robinson, M. (2014) "Animal personhood in Mi'kmaq perspective," *Societies* 4 (4): 672–88.
Robinson, M. (2017) "Intersectionality in Mi'kmaw and settler vegan values." In: Brueck, J. F. (ed.) *Veganism in an Oppressive World: A vegans-of-color community project*. Sanctuary Publishers, pp. 71–89.
Robinson, M. (2019) *Decolonizing body, mind, and spirit* [Virtual Conference given at Veganism of Color, 2019], 14 and 15 September.
Robinson-Jacobs, K. (2021) "Pandemic prompts more Black Americans to take up urban gardening to end food apartheid," *Mother Jones*, 19 November.
Rodriguez, S. (2018a) "Introduction." In: Rodriguez, S. (ed.) *Food Justice: A primer*. Sanctuary Publishers, pp. 7–29.

Rodriguez, S. (2018b) "Animal agriculture: An injustice to humans and nonhumans alike." In: Rodriguez, S. (ed.) *Food Justice: A primer*. Sanctuary Publishers, pp. 85–106.

Rohr, J. R., Barrett, C. B., Civitello, D. J., et al. (2019) "Emerging human infectious diseases and the links to global food production," *Nature Sustainability* 2 (6): 445–56.

Rushkoff, D. (2022) *Survival of the Richest: Escape fantasies of the tech billionaires*. New York: W.W. Norton.

Ryan, E. B., Fraser, D. & Weary, D. M. (2015) "Public attitudes to housing systems for pregnant pigs," *PLoS ONE* 10 (11): e0141878.

Ryder, R. D. (1975) *Victims of Science: The use of animals in research*. London: Davis-Poynter.

Ryder, R. D. (1989) *Animal Revolution: Changing attitudes towards speciesism*. Oxford: Basil Blackwell.

Saier Jr, M. H., Baird, S. M., Reddy, B. L. & Kopkowski, P. W. (2022) "Eating animal products, a common cause of human diseases," *Microbial Physiology* 32 (5–6): 146–57.

Sarkar, D., Walker-Swaney, J. & Shetty, K. (2020) "Food diversity and Indigenous food systems to combat diet-linked chronic diseases," *Current Developments in Nutrition* 4 (Suppl. 1): 3–11.

Satija, A. & Hu, F. B. (2018) "Plant-based diets and cardiovascular health," *Trends in Cardiovascular Medicine* 28 (7): 437–41.

Schlosser, E. (2005) *Fast Food Nation: The dark side of the all-American meal*. New York: Perennial.

Schor, J. B. (2011) *True Wealth: How and why millions of Americans are creating a time-rich, ecologically light, small-scale, high-satisfaction economy*. New York: Penguin Books.

Seaspiracy (2021) Directed by A. Tabrizi [Documentary]. A.U.M. Films & Disrupt Studios.

Seddon, N., Chausson, A., Berry, P., Girardin, C. A. J., Smith, A. & Turner, B. (2020) "Understanding the value and limits of nature-based solutions to climate change and other global challenges," *Philosophical Transactions of the Royal Society of London. Series B. Biological Sciences* 375 (1794): 20190120.

Shearer, J. K. (2018) "Euthanasia of cattle: Practical considerations and application," *Animals* 8 (4): 57.

Shields, S. & Greger, M. (2013) "Animal welfare and food safety aspects of confining broiler chickens to cages," *Animals* 3 (2): 386–400.

Shields, S., Shapiro, P. & Rowan, A. (2017) "A decade of progress toward ending the intensive confinement of farm animals in the United States," *Animals* 7 (5): 40.

Shiva, V. (2005) *Earth Democracy: Justice, sustainability and peace*. Cambridge, MA: South End Press.

Shiva, V. (ed.) (2016a) *Seed Sovereignty, Food Security: Women in the vanguard* of the fight against GMOs and corporate agriculture. Berkeley, CA: North Atlantic Books.

Shiva, V. (2016b) *Stolen Harvest: The hijacking of the global food supply*. Lexington, KY: University Press of Kentucky.

Shiva, V. (2016c) *Who Really Feeds the World?: The failures of agribusiness and the promise of agroecology*. Berkeley, CA: North Atlantic Books.

Shiva, V. (2016d) *The Violence of the Green Revolution: Third world agriculture, ecology and politics*. Lexington, KY: University Press of Kentucky.

Shiva, V. (2022) [Keynote Speech, Seed Fair, Florence].

Shivji, S. (2021) "Burdened by debt and unable to eke out a living, many farmers in India turn to suicide," *CBC News*, 30 March.

Simon, D. R. (2013) *Meatonomics*. San Francisco, CA: Red Wheel/Weiser.

Singer, M. (2010) "Ecosyndemics: Global warming and the coming plagues of the twenty-first century." In: Herring, A. & Swedlund, A. C. (eds.) *Plagues and Epidemics: Infected spaces past and present*. Oxford: Berg, pp. 21–37.

Singer, P. (1975) *Animal Liberation*. New York: Random House.

Singer, P., & Mason, J. (2007) *The Ethics of What We Eat: Why our food choices matter*. Emmaus, PA: Rodale Books.

Smith-Harris, T. (2004) "There's not enough room to swing a cat and there's no use flogging a dead horse: Language usage and human perceptions of other animals," *Revision* 27 (2): 12–15.

Sofos, J. N. (2008) "Challenges to meat safety in the 21st century," *Meat Science* 78 (1): 3–13.

Sorenson, J. (2010) *About Canada: Animal rights*. Halifax, NS: Fernwood Publishing.

Sorenson, J. (2016) *Constructing Ecoterrorism: Capitalism, speciesism & animal rights*. Halifax, NS: Fernwood Publishing.

Spalding, A. B. (2014) "Corruption, corporations, and the new human right," *Washington University Law Review* 91 (6): 1365–428.

Speciesism: The Movie (2019) Directed M. DeVries [Documentary]. United States: Mark DeVries Productions, Inc.

Statistics Canada (2019) *Overweight and obese adults, 2018*. https://www150.statcan.gc.ca/n1/pub/82-625-x/2019001/article/00005-eng.htm.

Statistics Canada (2023) *Hogs Statistics, Number of Farms Reporting and Average Number of Hogs Per Farm, Semi-Annual*, No. 32-10-0202-01 [Table]. https://www150.statcan.gc.ca/t1/tbl1/en/tv.action?pid=3210020201.

Sterling, S. R. & Bowen, S. A. (2019) "The potential for plant-based diets to promote health among Blacks living in the United States," *Nutrients* 11 (12): 2915.

Stibbe, A. (2001) "Language, power and the social construction of animals," *Society & Animals* 9 (2): 145–61.

Stop Ecocide Foundation (2021) *Independent Expert Panel for the Legal*

Definition of Ecocide: Commentary and core text. https://static1.squarespace.com/static/5ca2608ab914493c64ef1f6d/t/60d7479cf8e7e5461534dd07/1624721314430/SE+Foundation+Commentary+and+core+text+revised+%281%29.pdf.
Striffler, S. (2005) *Chicken: The dangerous transformation of America's favorite food*. New Haven, CT: Yale University Press.
Struthers Montford, K. & Wotherspoon, T. (2021) "The contagion of slow violence: The slaughterhouse and COVID-19," *Animal Studies Journal* 10 (1): 80–113.
Stucki, S. (2017) "(Certified) humane violence? Animal welfare labels, the ambivalence of humanizing the inhumane, and what international humanitarian law has to do with it," American Society of International Law. 111: 277–81.
Stull, D. D. & Broadway, M. J. (2004) *Slaughterhouse Blues: The meat and poultry industry in North America*. Belmont, CA: Wadsworth.
Szabo, Z., Koczka, V., Marosvolgyi, T., et al. (2021) "Possible biochemical processes underlying the positive health effects of plant-based diets – a narrative review," *Nutrients* 13 (8): 2593.
Tallberg, L. & Jordan, P. J. (2021) "Killing them 'softly' (!): Exploring work experiences in care-based animal dirty work," *Work, Employment and Society* 36 (5): 858–74.
Taylor, C. A., Boulos, C. & Almond, D. (2020) "Livestock plants and COVID-19 transmission," *Proceedings of the National Academy of Sciences of the United States of America* 117 (50): 31706–15.
Taylor, S. (2017) *Beasts of Burden: Animal and disability liberation*. New York: New Press.
The Economist (2014) "Capitalism and its critics: A modern Marx," *The Economist*, 3 May.
The Fair Housing Center of Greater Boston (n.d.) *1934–1968: FHA Mortgage Insurance Requirements Utilize Redlining.* https://www.bostonfairhousing.org/timeline/1934-1968-FHA-Redlining.html.
The New Corporation: The Unfortunately Necessary Sequel (2020) Directed by J. Bakan & J. Abbott [Documentary]. Toronto International Film Festival.
The White House (2021) *Report on the Impact of Climate Change on Migration.* https://www.whitehouse.gov/wp-content/uploads/2021/10/Report-on-the-Impact-of-Climate-Change-on-Migration.pdf.
Thomsen, P. T., Shearer, J. K. & Houe, H. (2023) "Prevalence of lameness in dairy cows," *The Veterinary Journal* 295: 105975.
Thu, K. (2009) "The centralization of food systems and political power," *Culture & Agriculture* 31 (1): 13–18.
Thu, K. (2010) "CAFOs are in everyone's backyard: Industrial agriculture, democracy, and the future." In: Imhoff, D. (ed.) *The CAFO Reader: The*

tragedy of industrial animal factories. Healdsburg, CA: Watershed Media, pp. 210–20.

Tilley, S. A. B. (2016) *Doing Respectful Research: Power, privilege and passion*. Halifax, NS: Fernwood.

Truth and Reconciliation Commission of Canada (2015) *Canada's Residential Schools: Missing children and unmarked burials – The final report of the truth and reconciliation commission of Canada*, Vol. 4, Montreal: McGill-Queen's University Press.

Tucker, J. A. (2020) *Epochs of ecology: The transition from feudalism to capitalism*. Master's Thesis, University of Denver.

Twine. R. (2012) "Revealing the 'animal-industrial complex' – A concept & methods for critical animal studies?," *Journal for Critical Animal Studies* 10 (1): 12–39.

UFCW Local 401 (2021) *Cargill High River Collective Agreement between Cargill Limited High River, Alberta & United Food and Commercial Workers Canada Union*. https://gounion.ca/wp-content/uploads/2022/12/Cargill-High-River-Searchable-CBA-Exp-December-2026.pdf.

UNESCO (2010) "Value of Water Research Report Series."

United Nations (1992) *Rio Declaration on Environment and Development*. Report of the United Nations Conference on Environment and Development. https://www.un.org/en/development/desa/population/migration/generalassembly/docs/globalcompact/A_CONF.151_26_Vol.I_Declaration.pdf.

United Nations (2009) *Human Rights: The right to adequate housing*. Fact Sheet 21, Rev. 1. https://www.ohchr.org/en/publications/fact-sheets/fact-sheet-no-21-rev-1-human-right-adequate-housing.

United Nations (2010) *The Right to Adequate Food*. New York: Office of the High Commissioner for Human Rights.

United Nations (2019) "Can we feed the world and ensure that no one goes hungry?," *UN News: Global Perspectives Human Stories*, 3 October.

United Nations (2022) *Biodiversity – Our strongest natural defense against climate change*. https://www.un.org/en/climatechange/science/climate-issues/biodiversity.

United Nations Environment Programme (2020) Preventing the next pandemic – Zoonotic diseases and how to break the chain of transmission. https://www.unep.org/resources/report/preventing-future-zoonotic-disease-outbreaks-protecting-environment-animals-and.

United Nations Environment Programme (2021) *Food Waste Index Report*. https://wedocs.unep.org/handle/20.500.11822/35280.

US Centers for Disease Control and Prevention (2023) Influenza Type A Viruses. https://www.cdc.gov/flu/avianflu/influenza-a-virus-subtypes.htm.

US Department of Agriculture (2008) Colostrum feeding and management on U.S. dairy operations, 1991–2007. https://www.aphis.usda.gov/animal_health/nahms/dairy/downloads/dairy07/Dairy07_is_Colostrum.pdf.

US Department of Agriculture (2016) *Organic Livestock Requirements.* https://www.ams.usda.gov/sites/default/files/media/Organic%20Livestock%20Requirements.pdf.

US Department of Agriculture (2023) *What Does Natural Meat and Poultry Mean?* https://ask.usda.gov/s/article/What-does-natural-meat-and-poultry-mean.

US Department of Agriculture (n.d.) What is "grass fed" meat? https://ask.usda.gov/s/article/What-is-grass-fed-meat.

US Department of Agriculture, Animal and Plant Health Inspection Service (2023) *2022–2023 Detections of Highly Pathogenic Avian Influenza.* https://www.aphis.usda.gov/aphis/ourfocus/animalhealth/animal-disease-information/avian/avian-influenza/2022-hpai.

US Department of Agriculture, Food Safety and Inspection Service (2013) *Veal from Farm to Table.* https://www.fsis.usda.gov/food-safety/safe-food-handling-and-preparation/meat/veal-farm-table.

US Food & Drug Administration (US FDA) (2020) *All About BSE (Mad Cow Disease).* https://www.fda.gov/animal-veterinary/animal-health-literacy/all-about-bse-mad-cow-disease.

US Geological Survey (n.d.) *What is the Difference between Low Pathogenic and Highly Pathogenic Avian Influenza?* https://www.usgs.gov/faqs/what-difference-between-low-pathogenic-and-highly-pathogenic-avian-influenza.

US House of Representatives (2022) "How the Trump administration helped the meatpacking industry block Pandemic worker protections," [Staff Report]. *Select Subcommittee on the Coronavirus Crisis Subcommittee.*

van den Hoonaard, D. K. & van den Scott, L. J. (2022) *Qualitative Research in Action: A Canadian primer*, 4th edn. Don Mills, ON: Oxford University Press.

Vasseur, E., Borderas, F., Cue, R. I., et al. (2010) "A survey of dairy calf management practices in Canada that affect animal welfare," *Journal of Dairy Science* 93 (3): 1307–16.

von Keyserlingk, M. A. G. & Weary, D. M. (2007) "Maternal behavior in cattle," *Hormones and Behavior* 52 (1): 106–13.

Water Footprint Network (n.d.) *Water Footprint of Crop and Animal Products: A Comparison.* https://www.waterfootprint.org/time-for-action/what-can-consumers-do/.

Watson, C. & Morse, F. (1976) *How Does it Feel to be a Tree?* New York: Parents' Magazine Press.

Weis, T. (2013) *The Ecological Hoofprint: The global burden of industrial livestock.* London: Zed Books.

Weis, T. (2018) "Ghosts and things: Agriculture and animal life," *Global Environmental Politics* 18 (2): 134–42.

Werner, C. & Osterbur, E. (2022) "Decoding plant-based and other popular diets: Ensuring patients are meeting their nutrient needs," *Physician Assistant Clinics* 7 (4): 615–28.

Wheeler, S. M. (2016) "Sustainability planning as paradigm change," *Urban Planning* 1 (3): 55–8.

Whiting, T. L. & Keane, M. A. (2022) "Animal protection and mass depopulation," *Canadian Veterinary Journal* 63 (8): 859–62.

Whiting, T. L. & Marion, C. R. (2011) "Perpetration-induced traumatic stress – A risk for veterinarians involved in the destruction of healthy animals," *Canadian Veterinary Journal* 52 (7): 794–6.

Wilkie, R. (2005) "Sentient commodities and productive paradoxes: The ambiguous nature of human–livestock relations in Northeast Scotland," *Journal of Rural Studies* 21 (2): 213–30.

Willett, W., Rockström, J., Loken, B., et al. (2019) "Food in the anthropocene: the EAT–Lancet Commission on healthy diets from sustainable food systems," *The Lancet* 393 (10170): 447–92.

Williams, K. C., Hernandez, E. H., Petrosky, A. R. & Page, R. A. (2009) "The business of lying," *Journal of Leadership, Accountability and Ethics* Winter: 11–30.

Wilson, E. (1984) *Biophilia: The human bond with other species*. Cambridge, MA: Harvard University Press.

Wingspread Conference (1998) Wingspread Statement on the Precautionary Principle [Academic conference at Wingspread, headquarters of the Johnson Foundation in Racine, Wisconsin], 23–25 January.

World Animal Protection (2020) *The Animal Protection Index*. https://api.worldanimalprotection.org/.

World Health Organization (2015a) *WHO Estimates of the Global Burden of Foodborne Diseases: Foodborne disease burden epidemiology reference group 2007–2015*. Geneva: WHO. https://apps.who.int/iris/handle/10665/199350.

World Health Organization (2015b) *WHO's first ever global estimates of foodborne diseases find children under 5 account for almost one third of deaths*. Geneva: WHO, 3 December. https://www.canada.ca/en/health-canada/services/food-safety-vulnerable-populations/food-safety-people-with-weakened-immune-system.html.

World Health Organization (2017) *WHO Guidelines on Use of Medically Important Antimicrobials in Food-Producing Animals*. Geneva: WHO. https://www.who.int/publications/i/item/9789241550130.

World Health Organization (n.d.) *Estimating the Burden of Foodborne Diseases*. https://www.who.int/activities/estimating-the-burden-of-foodborne-diseases.

World People's Conference on Climate Change (2010) *Cochabamba Declaration on the Rights of Mother Earth*. Cochabamba: Bolivia.

World Watch Institute (2016) *The State of Consumption Today.* http://www.adorngeo.com/uploads/2/7/3/5/27350967/the_state_of_consumption_today.pdf.

World Wildlife Fund (2020) *A Call to Stop the Next Pandemic.* https://c402277.ssl.cf1.rackcdn.com/publications/1348/files/original/FINAL_REPORT_EK-Rev_2X.pdf?1592404724.

World Wildlife Fund (2023) What is the sixth mass extinction and what can we do about it? https://www.worldwildlife.org/stories/what-is-the-sixth-mass-extinction-and-what-can-we-do-about-it.

Wrenn, C. L. (2015) "The role of professionalization regarding female exploitation in the nonhuman animal rights movement," *Journal of Gender Studies* 24 (2): 131–46.

Wrenn, C. L. (2019) "The Vegan Society and social movement professionalization, 1944–2017," *Food & Foodways* 27 (3): 190–210.

You, Y., Leahy, K., Resnick, C., Howard, T., Carroll, K. C., & Silbergeld, E. K. (2016) "Exposure to pathogens among workers in a poultry slaughter and processing plant," *American Journal of Industrial Medicine* 59 (6): 453–64.

Zaldua, J., Tanner, J., Jarrett, K., Gibbs, T., Bood, T. & Burbano, R. (2022) *Election Observation Report – Colombian Presidential Elections.* https://commonfrontiers.ca/wp-content/uploads/2022/08/COLOMBIAN-FRONTIERS-REPORT-2022-final.pdf.

Zerk, J. (2012) *Corporate Liability for Gross Human Rights Abuses: Towards a fairer and more effective system of domestic law remedies – A report prepared for the Office of the UN High Commissioner for Human Rights.* https://www.ohchr.org/sites/default/files/Documents/Issues/Business/DomesticLawRemedies/StudyDomesticeLawRemedies.pdf.

Index

Action Against Hunger 121
activism/social protest
   advocacy and 25–8
   assassination of deforestation activists 69–70
   and community building 192
   undercover investigations 159–60, 161–2, 164–6
   see also resistance
Adams, C. 91, 109, 111
African countries 14, 31, 65, 66
ag-gag laws 159–62
agroecology 150–1
Albritton, R. 122, 123
alcohol consumption: slaughterhouse workers 95
alienation, concept of 99–100
Amazon rainforest 69–70
"ambivalence" 174, 207
Ameli, K. 193–4
   biography 218
Animal Justice, Canada 160, 176–7
animal personhood 152
animal welfare issues *see* hidden world of industrial production
Animal-Industrial Complex (A-IC)
   confronting systems 55–64
   violence of 67–71
animals
   agency and resistance 112–15
   human and nonhuman 5, 6–7, 52
   recognizing as "beings with lives" 181–4
   recognizing labor of 104–8
Anthropocene 48–9
antibiotics and antibiotic resistance 137
Arluke, A. 80
   and Sanders, C. 174, 207
availability, accessibility, and adequacy of food 118–21
avian influenza ("bird flu"/H5N1/HPA1) 142–3

Balcombe, J. 6, 101–2, 163, 181, 182–3
Bartlett, C. et al. 195–6
Battiste, J. 189–90, 201–2
   biography 218–19
Battiste, M. 24–5
Baur, G. 113–14, 155, 204–5, 207
   biography 219
bearing witness 109, 110, 206
beef cattle 69, 70
"beings with lives", recognizing animals as 181–4
Bekoff, M. 173–4, 178, 182, 183, 207
   biography 219–20
Belasco, W. 151
Bell, M. M. and Ashwood, L. 122, 178
Berry, W. 119–20, 151
BIOS research unit, Finland 203
"bird flu" (H5N1/HPA1) 142–3
Black/African Americans 89, 124, 125–6, 129–31
Blanchette, A. 104, 111, 112
Blattner, C. 141, 145
   et al. 76, 105, 106, 112, 172
blocking out 207, 208–9
Bohanec, H. 179–80
boundary work/"boundary labour" 77, 80
#Boycottmeat movement 112
Brueck, J. F. 134
BSE 139

cage-free designation 176
Campbell, T. C. and Campbell, T. M. 71, 123
Canada
   Animal Justice 160, 176–7
   farm sizes 82
   Food Guide 149–50, 200
   Health of Animals Act (2019) 169
   Health Canada 138, 139, 149–50
   NFACC Codes of Practice 81, 82–3, 166
   Truth and Reconciliation Commission 2
   see also entries beginning Indigenous

Index

Canadian Broadcasting Corporation (CBC) 78, 165
Canadian Criminal Code (RSC) 166–7
Canadian Food Inspection Agency (CFIA) 167–9, 171–2
capitalism
  climate change and infectious diseases 144
  dreams and nightmares 47–75
  logic of 30–2, 78–86, 201
  as silent ingredient 134
  see also colonial-capitalist diet; neoliberalism
Cargill 53, 73, 88, 89
"caring-killing" paradox 80
Carleton, E. 14
Carrington, S. 129
cattle *see* beef cattle; dairy industry
Ceballos, G. et al. 17
certified organic label 176, 177
"cheap food regimes" 117–18
chickens 83, 100, 104–5, 169
China, US and EU emissions 75
Chomsky, A. 35, 36, 38, 47, 65–6, 68, 72, 75, 150, 178, 199–200
  biography 220–1
Chomsky, N. 60
Clark, M. A. et al. 64
climate change
  denial 47–50, 52, 55–6
  as global imperative 187
  impacts of agriculture 67–71
  impacts in Global South 73–4
  and infectious diseases 144
  and relationship to land 16–17, 65–7
  social protests 25–8
  see also GHGs/emissions
Climate Change Task Force (CCTF) 25–6, 27
Collins, P. H. and Bilge, S. 7
Colombia 36, 39–42
colonial-capitalist diet 117–18
  deconstructing "ideal" body 123–6
  food insecurity to food justice 128–35
  incremental shift to compassionate food future 146–51
  overconsumption and underconsumption issues 121–3
  politics and resistance 151–3
  right to food 118–21
  transparency and beyond 153–4
  zoonotic diseases 135–46, 147
colonialism
  consequences of 56–8
  decolonized worldview 190
  and neocolonialism 65–6, 67
  see also under Indigenous Peoples
community building 191–2
  and social infrastructure 126–8

compartmentalization and de-animalization 90–2
compassion ("Big C")
  response to structural violence 21–5
  towards radical democracy based on 34–8
compassionate food system 185–6, 209–12
  emergent themes 188–98
  incremental shift to 146–51
  joy and gratitude 204–9
  short- and long-term politics 198–200
  see also democracy/radical democracy; precautionary principle; transparency
concentrated animal feeding operations (CAFCs) 146–7
confronting systems 55–64
consumer-citizens 119–20, 134–5, 161, 165, 174–5
"consumer lock-in" 126–7
consumerism and individualism 126–8
consumption, over and under 121–3
COP27, UN 63, 67, 74–5
corporation concentration 81–2, 88–9, 97, 103
  reduction and impact 146–7
corporations
  dependency on 33–4
  lying 61
  meatpacking 84, 86–9
  and small farmers 53–5
Coulter, K. 106–7
COVID-19 pandemic 32
  meat processing workers 86–8, 92–3
  raised awareness of zoonotic diseases 141, 142, 143–4, 145
  slaughterhouse and farm workers 78–86, 114–15
"cowboy capitalism" 59, 118
coyotes, San Francisco 208–9
Crenshaw, K. 62
Creutzfeldt-Jakob disease 139
Critical Animal Studies (CAS) 124
crowding and stress 142
Cuba 36
culture shock 158–9

dairy industry 101, 105, 167, 169–72
  and veal industry 172–3
Dalton, J. 164, 165, 166
Davis, K. 100, 104–5
de-animalization, compartmentalization of human work and 90–2
decolonized worldview 190
deforestation 68–70
degrowth 66–7, 192
"Democracy Index" (EIU) 35–6, 37
democracy/radical democracy 34–8, 186–8, 202–4
  Earth and Species 18–21
  see also precautionary principle

denial 47–50, 52, 55–6
depopulation of animals: COVID-19 pandemic 79–83
deskilling and "entanglements of oppression" 99–104
"development", neoliberalism and myth of 38–42
DeVries, M. 5
  biography 220
dirty work and slow violence 92–3, 97–8
diversity 190, 192–3
  and individuals 191
doublethink 174, 207
Dryden, J. and Rieger, S. 78, 86, 88

Earth Day (1970) 185
Earth and Species democracy 18–21
"ecocide": ICC crime and definition 28–9
E. coli 136, 137, 139
ecology-versus-economy arguments 203
Economic Intelligence Unit (EIU): "Democracy Index" 35–6, 37
The Economist 32
ecosystem services 17, 56
Ecuador Constitution 29–30
education 148, 153–4, 200
  decolonization of 24–5
  "popular education" model 25
effective change, initiating 192–8
Einstein, A. 20, 196
elites 52, 59–60, 62
Ellis, C. 80
emissions *see* GHGs/emissions
emotions
  and intelligence of farmed animals 173–4, 182–3
  joy and gratitude 204–9
  see also mental health
empathy 191
  cultivating 108–10
Enclosure Acts, Britain 23
"entangled empathy" 109
ethics of care 109
"Etuaptmumk"/"Two-Eyed Seeing" (E/TES) 195–6
European Institute for Animal Law and Policy 177
European Union (EU)
  animal transport and related operations 169
  China and US emission 75
euthanasia *see* depopulation of animals

Farm Sanctuary 113–14, 207
farmers' mental health 79–80, 81
  and suicides 53, 79
farms
  depopulation of animals: COVID-19 pandemic 79–82

multinational corporations and small farmers 53–5
sizes 13–14, 81–2, 119, 146
fat phobia 125–6
fertilizers and pesticides 72
Finland: BIOS research unit 203
Fischer, F. 49
fishes 6, 101–2, 137, 163–4
Fitzgerald, A. J. 7, 9, 98, 104, 138, 142, 160, 166, 167–8, 177
  biography 220–1
Fitzgerald, A. J. et al. 99
Foer, J. S. 150
Food and Agriculture Organization (FAO) 12, 13–14, 26, 150–1, 163
"food deserts"/"food apartheid" 129, 130
Food Guide, Canada 149–50, 200
food insecurity
  to food justice 128–35
  see also hunger; inequalities
food sovereignty 65–6, 151–3, 210
Food Tank, US 135
food waste 12–13
foodborne and prion diseases 136–41
fossil fuel dependency 74–5
Foster, J. B. and Burkett, P. 23
free-range designation 176
Freire, P. 25
future directions 45–6
  see also compassionate food system
future generations 42–5
  Seven Generations principle 189

Galtung, J. 15, 45
gestation and farrowing crates 100, 165
GHGs/emissions 75
  Africa 14
  agriculture and food waste 13
  animal agriculture 64, 67–8
Gibbs, T. 22, 37, 44, 120, 156, 196
Gibbs Leech, O. 26–7, 43–4
  biography 221
Gillespie, K. 7, 109–10, 170, 171, 174–6, 177, 207
  and Lopez, P. 47
Global North/Global South relationship 209–12
  Amazon rainforest 69–70
  COP27 74–5
  global democracy 35–8
  Green Revolution 72
  impacts of climate change 73–4
  inequalities 14, 15, 31–2, 45
  neoliberalism and myth of "development" 38–42
Gorz, A. 66
government legislation *see* labeling; legislation and regulation

Graf, A. 45
Grandin, T. 179
grass-fed designation 176
gratitude and joy 204–9
Gray, A. 170–1
Green New Deal, US 199–200
Green Revolution
  proponents 160
  recovering from 71–4
Greenebaum, J. 124, 134
  biography 221
greenwashing and humane washing 178
Greger, M. 142, 145
Gruen, L. 109
Guterres, A. 27–8, 63, 67

Haedicke, M. A. 84, 85–6, 92–3, 108–9
  biography 221–2
Harper, A. B. 4–5, 25, 125–6, 130, 131
  biography 222
Harper, C. L. and Le Beau, B. F. 119
Harwatt, H. 64
Head, L. 48–9
health
  and demographics of workers 86–90
  diet-related diseases 71, 122–3, 128–9, 131, 147
  Indigenous Peoples 131–2
  mental 79–80, 81, 93–8
  see also zoonotic diseases
Health of Animals Act (2019), Canada 169
Health Canada 138, 139, 149–50
Healthy Food Access Programs, US 129
Heredia, N. and García, S. 136
Herring, D. A. and Swedlund, A. C. 144
hidden world of industrial production 155–6
  humane production labels and standards 175–81
  individual abuses and standard industry practices 163–75
  recognizing nonhuman animals as "beings with lives" 181–4
  speciesism in action 156–8
  why we don't know what we don't know 158–62
Hoekstra, A. Y. 70–1
Holt-Giménez, E. 72, 76, 121–2, 134
human rights
  availability, accessibility and quality of food 118–21
  corporate violations 61
  future generations 42–5
  protection, peace, and food justice relationship 50–5
  and responsibilities 33–4
Human Rights Watch 40
humane production claims 175–81
humane washing 178–9

Hume, S. 102
humility 190, 196
hunger 13–14, 65, 67
  and food waste 12–13
  underconsumption and overconsumption 121–3
Hurricane Fiona 16, 25, 127–8, 204

"ideal" body, deconstructing 123–6
immigrant (illegal/undocumented) workers 84–5, 89, 93
India
  agricultural corporations and farmers 53, 54–5
  Navdanya 19, 73, 210
Indigenous knowledge 1–2, 18, 20, 24–5, 43–4, 153, 195–6
Indigenous Northern Affairs Committee study 189–90
Indigenous Peoples
  colonization 56–8
  and decolonization of education 22–5
  diet and health 131–2
  food sovereignty 151–3
  impact of Hurricane Fiona 128
  inequalities 15–16
  perspective on initiating change 193
  relationship with the land 187, 195, 196
  Seven Generations principle 189
individual abuses and standard industry practices 163–75
individualism and consumerism 126–8
individuals and diverse populations 191
industrial food production 6–7, 119–21
  see also Animal-Industrial Complex (A-IC); Green Revolution; hidden world of industrial production; labor in industrial production; zoonotic diseases
inequalities
  Global North/Global South 14, 15, 31–2, 45
  Indigenous and racialized groups 15–16
  over and under consumption 121–3
influenza 141–6
Inoue, T. 163–4
  biography 222
institutional settings, plant-based diet in 146
intellectual property rights 54
intelligence and emotions of farmed animals 173–4, 182–3
interconnection/interdependence 20–1, 22
  failure to recognize 120–1
  as framework for advocacy and social protest 25–8
  initiating effective change 192–8
  "sustainable happiness" 200
  "Three Sisters" story 153

interdisciplinary approach 3–4, 7, 194–5, 203
International Livestock Research Institute (ILRI) 14
International Monetary Fund (IMF): SAPs 38–9
intersectional approach 3–4, 7, 25, 62, 199
  and cross-movement solidarity 188
  food justice 132–3, 134
  "ideal" body 124–5
  "logic" of capitalism 201
  meatpacking and slaughterhouse workers 86, 89

Jacques, J. R. 98–9
Jamail, D. 69–70, 206, 209, 211
  and Bendell, J. 62–3
Joy, M. 159, 161
joy and gratitude 204–9
junk/fast food 122–3, 131, 147

Kabir, S. M. L. 138
Kateman, B. 148
Kavanagh, L. 149
  biography 222–3
Keim, B. 106, 107, 115, 208–9
  biography 223
Kelly, W. 185
Keshen, R. 155
  biography 223
Kevany, K. 146
Kimmerer, R. W. 18, 24, 30, 43–4, 153, 202, 206
King, B. J. 181
Klinenberg, E. 126, 127–8
knockers: role and emotional impact 94–5
Ko, A. and Ko, S. 124
Kopecky, A. 160
Kumar, K. and Makarova, E. 126

Labchuk, C. 160
labeling
  humane production claims 175–81
  sustainability 201–2
labor in industrial production 76–7
  compartmentalization of human work and de-animalization 90–2
  COVID-19 pandemic 78–86, 114–15
  deskilling and "entanglements of oppression" 99–104
  dirty work and slow violence 92–3, 97–8
  health and demographics of workers 86–90
  mental health impacts 79–80, 81, 93–8
  recognizing animal labor 104–8
  resistance and beginning of just transition 110–16
  spillover effect 98–9

transparency and beyond: cultivating empathy 108–10
Ladd, A. E. and Edward, B. 90
land
  and climate change relationship 16–17, 65–7
  "commons" 23, 119
  degradation and environmental pollution 68
  and democracy 187
  dominion over 65–7
  Indigenous Peoples' relationship with 187, 195, 196
  interconnection/interdependence 120–1, 153
  "Land Ethic" 20–1
  laws protecting nature 28–30
Land, B. 129
Leech, G. 31, 36, 45–6, 53, 54, 87, 91–2
Leech, O. see Gibbs Leech, O.
legislation and regulation
  ag-gag laws 159–62
  animal welfare 167–9, 208
  banning intensive farming 146–7
  land and other species relationship in 28–30
  livestock transport 161–2, 168–9, 171–2
  see also labeling
Leopold, A. 20–1
livestock transport 161–2, 167–9, 171–2
Long, A. 193
  biography 223
Lorenzen, J. A. 126–7
Louv, R. 200
Lymbery, P. 145

McBay, A. 132
  biography 224–5
MacCath-Moran, C. 110, 124–5, 161–2
  biography 224
McGranahan, C. 111
Marshall, A. 18, 23–4, 27, 28, 33–4, 44, 145–6, 185, 187, 195, 202, 212
  biography 224
Martin, A. 135
Martin, J. et al. 51
Marx, K. 99–100
Mason, J. 181–2
Masson, J. M. 170, 173, 182, 183
May, E. 34
meat eating/carnism 122, 159
meatpacking industry 84, 86–9
Mediterranean diet 148
mental health 79–80, 81, 93–8
Mesquita, R. 69, 70
Mi'kma'ki 1
Monsanto 53, 54, 73

Montenegro de Wit, M. 142
Mora, D. C. et al. 93
mosaic metaphor of initiating change 194
Mosby, I. and Rotz, S. 89
"Msit No'kmaq" ("All My Relations") 195, 198
multispecies ethnography 194–5
mutuality 204–5, 206

National Farm Animal Care Council (NFACC) Codes of Practice, Canada 81, 82–3, 166
National Wildlife Federation (NWF), US 17
"Nature Deficit Disorder" 200
nature-based solutions (NBS) 50–1
Navdanya, India 19, 73, 210
"needs" vs "wants" 134–5
neoliberalism 35, 120
  and loss of community 126–8
  and myth of "development" 38–42
"Netukulimk", concept of 195, 196
Nibert, D. 6, 7, 56–7, 59, 61, 62, 103, 156–7, 158
  biography 225
Nixon, R. 97
Norgaard, K. M. 49
Norwood, F. B. and Lusk, J. L. 180
Noske, B. 99–100

obesity 122–3
O'Brien, C. 200
  biography 225
Okorie-Kanu, E. et al. 138
overconsumption and underconsumption issues 121–3
Oxfam International 32, 117

Pachirat, T. 94–5
paradigm shift 188–91
Parish, J. 79, 119, 143–4, 153
  biography 225–6
Patel, R. 121
  and Moore, J. 84–5, 104, 117–18
"perpetration-induced traumatic stress" (PITS) 94
pesticides and fertilizers 72
Pew Commission on Industrial Farm Animal Production 140–1, 142
pigs/hogs 78, 80–2, 83
Piketty, T. 32
plant-based diet 130–1, 133
  benefits and promotion of 147–51
  education 200
  institutional settings 146
Plant-Based Treaty initiative 74–5
plants
  bacterial contamination 139
  Indigenous knowledge 153

play, in nonhuman animals 182–3
politics
  short and long term 198–200
  see also activism/social protest; resistance
Pollan, M. 151
Poore, J. and Nemecek, T. 63–4
Porcher, J. 99
precautionary principle 28, 145–6, 187–8, 198, 202–4
prion diseases 139–40
processed and junk/fast food 122–3, 131, 147
PROOF 15

quitting: slaughterhouse workers 111

radical democracy see democracy/radical democracy
"redlining" 129–30
'reducetarian' diet 148
refusal 111
regulation see labeling; legislation and regulation
research
  authors' approach and background 7–10
  methods 213–17
  participant biographies 218–28
resistance
  food as politics 151–3
  worker strategies and beginning of just transition 110–16
  see also activism/social protest
responsibilities and rights 33–4
Ritchie, H. and Roser, M. 68, 69
rights see human rights
Riley, S. 34, 177–8, 208
  biography 226
Ritzer, G. 131
Robinson, M. 9, 107, 108, 117, 125, 131–2, 134, 146, 151–3
  biography 226
Rodriguez, S. 133
Rohr, J. R. et al. 137, 140, 141
Rooke, P. 76, 95–6
  biography 226
Rushkoff, D. 47, 48, 52, 62
Ryan, E. B. et al. 180
Ryder, R. D. 157, 158

Saier Jr, M. H. et al. 163
Sainath, P. 54
Salmonella 137, 138, 139
Sarkar, D. et al. 153
Satija, A. and Hu, F. B. 148
Sawant, K. 47–8
Scandinavian societies 203
Seddon, N. et al. 50–1
seed banks: Navdanya, India 73, 210

seed corporations 53–4
Seed Fair, Italy (2022) 72–3
Seven Generations principle 189
Seven Sacred Gifts of Life 196
Shearer, J. K. 81
Shields, S. and Greger, M. 138
Shiva, V. 12, 18, 19–20, 21, 30, 43, 53, 59, 72–3, 121, 133, 144, 154, 190, 202, 210, 212
  biography 227
Simon, D. R. 122
Singer, P. 144, 158
  and Mason, J. 134–5
Singh, A. 54–5, 120–1, 133
  biography 228
slaughterhouses *see* hidden world of industrial production; labor in industrial production
slow violence 92–3, 97–8
social infrastructure 126–8
Sorenson, J. 160
soy production 68, 69
Spalding, A. 60
species destruction 16–17
speciesism 156–8
spider web metaphor of initiating change 193
spillover effect of slaughterhouse work 98–9
standard industry practices
  and individual abuses 163–75
  and organic certification 177
Sterling, S. R. and Bowen, S. A. 130, 148
Stop Ecocide Foundation 28–9
stress and crowding 142
Structural Adjustment Programs (SAPs) 38–9
structural violence 15, 67, 208, 212
  access to food and housing 129–30
  and logic of capitalism 30–2, 201
  responding through compassion 21–5
Struthers Montfort, K. and Wotherspoon, T. 97, 112
Stucki, S. 175
sustainability 150
  labeling 201–2
"sustainable happiness" 200

Taylor, C. A. et al. 78–9, 86
Taylor, S. 181
Thich Nhat Han 20
Thoreau, H.D. 155, 156
"Three Sisters" story 153
Thu, K. 119, 120
Thunberg, G. 26, 42
transparency 9–10, 153–4, 156, 200–2
  cultivating empathy 108–10
  fisheries 163–4
  paradigm shift 190–1

transport of livestock 161–2, 167–9, 171–2
Truth and Reconciliation Commission of Canada 2
Tutu, D. 20
Twine, R. 2, 55, 67, 163
Two-Eyed Seeing approach 195–6

Unama'ki (Cape Breton Island) 16
undercover investigations 159–60, 161–2, 164–6
United Nations (UN) 27–8, 61, 63
  Climate Action 50
  COP27 63, 67, 74–5
  food safety and zoonotic diseases 135, 144–5
  hunger and overconsumption 121
  right to food 118
  Rio Declaration: Precautionary Principle 28
  Security Council 51
  UNEP 12–13, 14, 144–5
  UNESCO 70
  UNICEF 65
  see also Food and Agriculture Organization (FAO); World Health Organization (WHO)
United States (US)
  Black/African Americans 89, 124, 125–6, 129–31
  Centers for Disease Control and Prevention 143
  Department of Agriculture (USDA) 87, 164, 170, 172, 176
  EU and China emissions 75
  farm sizes 81–2
  Farm System Reform Act 146
  Food Tank 135
  Geological Survey 142
  Healthy Food Access Programs 129
  House of Representatives 86–8, 92
  Humane Methods of Slaughter Act (1978) 166
  National Wildlife Federation (NWF) 17
  sugar industry 59
universal basic income (UBI) 112

Vassallo, L. 192
  biography 228
veal industry 172
veganism 124, 125, 148
vegans of color 124, 134, 152
vegetarianism 148
veterinarians 79–80, 81
violence
  of Animal-Industrial Complex (A-IC) 67–71
  crime as spillover effect 98–9
  slow 92–3, 97–8

von Keyserlingk, M. A. G. and Weary, D. M. 170

Wadiwel, D. 172
Waring, M. 184
Washington, K. 129, 130
water conflict 51
water footprint and Water Footprint Network 70–1
Watson, C. 18
Weis, T. 69
Werner, C. and Osterbur, E. 148
Wheeler, S. M. 150
Willett, W. et al. 147
Williams, K. C. et al. 61
Wilson, E. 20

Wingspread Statement on the Precautionary Principle 28
World Animal Protection Organization 149, 166–7, 169
World Bank 32, 72
World Day for Farmed Animals (WDFA) 155
World Food Program 13–14, 65
World Food Summit 13
World Health Organization (WHO) 59, 136–7
World People's Conference on Climate Change 29
World Watch Institute 122
World Wildlife Fund 17, 141

Zerk, J. 61
zoonotic diseases 135–46, 147